DISCRETE CONTACT MECHANICS WITH APPLICATIONS IN TRIBOLOGY

Elsevier Series on Tribology and Surface Engineering

DISCRETE CONTACT MECHANICS WITH APPLICATIONS IN TRIBOLOGY

IRINA GORYACHEVA
Ishlinsky Institute for Problems in Mechanics of the Russian Academy of Sciences, Moscow, Russia

YULIA MAKHOVSKAYA
Ishlinsky Institute for Problems in Mechanics of the Russian Academy of Sciences, Moscow, Russia

Elsevier
Radarweg 29, PO Box 211, 1000 AE Amsterdam, Netherlands
The Boulevard, Langford Lane, Kidlington, Oxford OX5 1GB, United Kingdom
50 Hampshire Street, 5th Floor, Cambridge, MA 02139, United States

ISBN: 978-0-12-821799-3

For information on all Elsevier publications visit our website at https://www.elsevier.com/books-and-journals

Publisher: Matthew Deans
Acquisitions Editor: Dennis McGonagle
Editorial Project Manager: Catherine Costello
Production Project Manager: Kamesh Ramajogi
Cover Designer: Vicky Pearson Esser

Typeset by TNQ Technologies

Contents

Preface

Tribology is both a fundamental and an applied science aimed to explain the phenomena of friction and wear of materials in contact interaction as well as to control these processes based on the deep understanding of them. Tribology development is based on both the theoretical and experimental study.

The development of fundamental tribology is strongly connected with the progress in contact mechanics and fracture mechanics as well as physics and chemistry. Mechanics contributes significantly to the study of friction and fracture of surfaces, since all processes of various natures occurring during friction in the surface layers of contacting bodies are greatly influenced by the magnitude of stresses acting there. High real pressures and sliding velocities at and near contact spots cause significant temperature rise in these areas, which changes the properties of subsurface layers. They also cause mechanical and temperature stresses there, which facilitate chemical reactions and activate interdiffusion. In the surface layer undergoing high deformations, crack initiation and development occur, which results in fracture (wear) of the material.

Explanation of the mechanisms of friction and wear of materials under various contact conditions, which is one of the key issues of fundamental tribology, is impossible without formulating and solving contact problems taking into account not only the macrogeometry of interacting surfaces, but also their microgeometry.

A surface microrelief in the form of waviness and roughness causes a discrete contact region. As a result, high pressures arise on the real contact spots, leading to the stress concentration in a thin surface layer, the thickness of which is comparable with the characteristic size of a contact spot. The distribution of stresses in this layer largely determines the nature of its fracture, i.e., its surface wear.

The characteristics of contact interaction and the magnitude of the friction force are influenced by both the parameters of the surface microrelief of interacting bodies and their mechanical characteristics, as well as by the physical properties of the surfaces and the medium between them. Adhesive attraction between the surfaces, caused by intermolecular forces and capillary pressure in liquid bridges, leads to a redistribution of contact pressure at the microscopic level and increases the real contact area, which significantly affects the friction and wear behavior. The combined effect of the surface microrelief and adhesion leads to adhesion hysteresis and irreversibility of the process of formation and destruction of contact spots, which is one of the main mechanisms of friction force (its adhesive component).

Another important contribution to the friction force is the deformation (hysteresis) component associated with the energy loss that occurs when surface layers of an imperfectly elastic material are deformed by asperities of a counterbody during relative motion of the surfaces. The microrelief and discreteness of contact play an important role in this mechanism of friction: they, together with the velocity of relative motion of the surfaces, define the frequency and amplitude of material deformation, which determine the hysteresis losses.

In this book, models of contact interaction are presented, which are developed to study the distribution of contact stresses in normal and tangential contacts of elastic and viscoelastic bodies with given macroshapes of the contacting surfaces, also taking into account a surface relief (waviness, roughness) at microscale. A distinctive feature of all the developed models is the use of analytical methods based on solving mixed problems of contact mechanics for a system of contact spots, taking into account their mutual influence. The results obtained allow one to analyze the influence of both the shape of asperities and density of their location on the contact characteristics (distribution of real contact pressure, real contact area, approach of bodies at given values of nominal pressure, etc.) in normal approach and retraction of the surfaces, as well as on the friction force in the relative contact motion.

In the first three chapters, models are developed for studying the normal contact of elastic bodies taking into account their surface microgeometry. Chapter 1 studies the normal approach of elastic bodies with nominally flat surfaces having a regular microrelief. An approximate method for solving periodic contact problems based on the principle of localization is proposed. By using this method, it is possible to analyze the influence of not only the shape of an individual asperity, but also the density of their location on the real contact area and additional displacement of elastic bodies due to the presence of a rough layer on their surfaces. This method is used in Chapter 2 to solve contact problems for elastic bodies with nominally flat surfaces, taking into account their microgeometry and adhesive interactions of a different nature (molecular and capillary adhesion). In addition, in Chapter 1, an approximate analytical solution of the problem of indentation of a bounded system of asperities into an elastic half-space is constructed, and the role of the edge effect on the distribution of forces between the asperities is studied, taking into account their mutual influence. The additional displacement function associated with the presence of a rough layer on the surface is analyzed in detail in Chapter 3, including the effect of the adhesive forces between the surfaces. Taking into account the additional displacement function, the formulation and the method of solution of contact problems for bodies of a given macro- and microgeometry are also presented in Chapter 3. The obtained solutions of the model problems made it possible to analyze the influence of the surface roughness parameters (shape, density and height distribution of asperities) on the nominal contact pressure distribution at macrolevel, as well as on the size of the nominal contact area.

Chapters 4 and 5 are devoted to modelling the adhesive and deformation (hysteresis) components, respectively, of the friction force in the sliding contact of deformable bodies. In Chapter 4, an analytical model is developed for calculating the energy loss in an approach-retraction cycle of two asperities, which is used to calculate the adhesive component of the friction force during relative sliding of two elastic half-spaces, whose surfaces are covered with asperities of a given shape. The dependence of the adhesive friction force in sliding contact on the parameters of microgeometry, adhesion, and elastic properties of the contacting bodies is analyzed. The adhesive component of the rolling friction force is also calculated, and its dependence on the height distribution of asperities on the surface of the rolling body is studied.

In Chapter 5, a model is developed for calculating the friction force caused by hysteresis losses during cyclic deformation of the surface layer of an imperfectly elastic body when a rough counterbody slides over it. Based on the obtained analytical and numerical-analytical solutions of discrete contact problems for a viscoelastic half-space (or viscoelastic layer) under conditions of uniform sliding of a surface with microgeometry over it, the influence of the microgeometry parameters (roughness shape, density of location), viscoelastic characteristics of the body, and sliding velocity on the contact characteristics and hysteresis component of the friction force is studied. This approach is used to analyze the combined effect of micro-geometry and adhesive forces, as well as liquid contained in the gap between the surfaces on the hysteresis component of the friction force.

Thus, the developed analytical models make it possible to study the effect of surface microgeometry on the contact characteristics for various types of interaction of deformable bodies (normal approach, relative sliding, and rolling), and for various contact conditions (dry and lubricated surfaces, liquid in the gap, etc.), as well as to analyze the friction force as a function of the mechanical and surface properties of the contacting bodies and their microgeometry.

These studies were carried out in the Ishlinsky Institute for Problems in Mechanics of the Russian Academy of Sciences partly under the financial support of the Russian Foundation for Basic Research (grant 20-01—00400).

CHAPTER 1

Normal discrete contact of elastic solids

The classical contact problem is usually formulated for topographically smooth contacting surfaces, so the contact region is continuous. But it is well known in tribology that a real contact region consists of contact spots distributed within the nominal (average) contact region. So, the real contact area is a small fraction of the nominal contact area.

The main reason of the contact region discreteness is the roughness and waviness of the contacting surfaces. In the general case, surface topography is represented by a combination of deterministic and random functions (Whitehouse, 1994) determined by natural factors or technological treatment of the surface. Deterministic components are formed either as a result of imperfections in the operation of technological equipment or in stationary operating conditions, for example, the steady shape of a worn surface (Goryacheva, 1998). In addition, a regular microgeometry on the surface can be created to control the operational properties of friction pairs, in particular their tribological characteristics (Goryacheva and Tsukanov, 2020a).

Note that the discrete contact problem also arises in study of the contact interaction of composite materials and inhomogeneous bodies with inclusions. The contact discreteness can be also formed by a special surface treatment, due to structural inhomogeneity, by using discontinuous coatings, and so on.

In this chapter, we present the approaches to study the contact characteristics of a system of the finite number of asperities penetrating into the elastic half-space and the characteristics of an elastic contact of nominally flat surfaces (at macro level), one of which has a regular microgeometry.

1.1 Formulation of the discrete contact problem

We give here the general formulation of the discrete contact problem. Let us consider a contact interaction of a deformable half-space and a counter

Discrete Contact Mechanics with Applications in Tribology
ISBN 978-0-12-821799-3
https://doi.org/10.1016/B978-0-12-821799-3.00002-9

body, the shape of which is described by the function $z = -F(x, y)$ in the system of coordinates connected with the half-space (the plane Oxy coincides with the half-space surface in the undeformed state, and the z-axis is directed into the half-space). After deformation, a finite number N or an infinite number of contact spots ω_i occur at the surface $z = 0$ of the half-space within the nominal contact region Ω (Fig. 1.1). If $N \to \infty$, the region Ω coincides with the plane $z = 0$.

The real contact pressure $p_i(x, y)$ acts at each contact spot $(x, y) \in \omega_i$. We assume here that shear stresses are negligibly small. The contact pressure provides the displacement of the half-space surface along the z-axis. This displacement $u_z(x, y)$ depends on the pressures $p_i(x, y)$ $(i = 1, 2, \ldots N)$ applied to all contact spots:

$$u_z = A[p_1, p_2, \ldots, p_N] \tag{1.1}$$

The operator A is determined by the model of the deformable bodies in contact. For the contact between a rigid body with a rough surface and an elastic half-space, the relation is

$$u_z(x, y) = \frac{1 - \nu^2}{\pi E} \sum_{i=1}^{N} \int\int_{\omega_i} \frac{p_i(x', y')dx'dy'}{\sqrt{(x' - x)^2 + (y' - y)^2}} \tag{1.2}$$

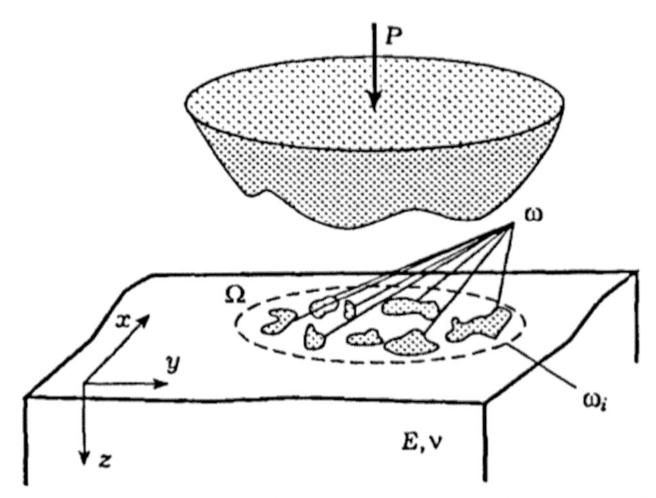

Figure 1.1 Scheme of the contact of a rigid body with rough surface and the elastic half-space.

Here, E and ν are the Young's modulus and Poisson's ratio of the half-space, respectively.

The contact condition at each contact spot ω_i is

$$u_z(x,y) = D - F(x,y), \quad (x,y) \in \omega_i, \tag{1.3}$$

where D is the displacement of the rigid body along the z-axis. If D is not given in advance, but the total load P applied to the bodies and directed along the z-axis is known, the equilibrium equation must be added to Eqs. (1.1) and (1.3):

$$\sum_{i=1}^{N} \int\int_{\omega_i} p_i(x,y)\,dxdy = P \tag{1.4}$$

The system of Eqs. (1.2)−(1.4) can be used to determine the real contact pressure $p_i(x,y)$ within the contact spots ω_i. However, the solution of this contact problem is very complicated, even if the sizes and the arrangement of contact spots are known. In the general case, we must determine also the number N, and the positions and shapes of the contact spots ω_i for any value of the load P. For a differentiable function $F(x,y)$, we can use the condition $p_i(x,y)|_{x,y\in\partial\omega_i} = 0$ to determine the region ω_i of an individual contact. Note that the contact problem with unknown contact regions ω_i is nonlinear, so it is not possible to use the superposition method for its solution.

One of the methods developed to solve the discrete problem is the use of the known exact solutions for the complete contact within the nominal contact region Ω and then the step-by-step modification of the contact region to get the positive contact pressures within the real contact spots ω_i and zero pressures outside them (Chekina and Keer, 1999). This method is based on the exact analytical solutions of the contact problems in 2-D (Muskhelishvili, 1953) and 3-D formulations (Galin, 1953, 2008).

The contact problem formulated above can be solved numerically based on the geometric description of the contacting surfaces obtained from measurements (Lubrecht, 1991; Stanley, 1997). In this case the faithfulness of the stress-strain state so determined depends on the accuracy of the numerical procedure.

It is worth noting that there is little point in developing an exact solution of the discrete contact problem because the function $F(x,y)$ is usually determined approximately by measurements of some small surface element before deformation. There are basic constraints on the accuracy of measurements of a surface microgeometry by different devices. The function $F(x, y)$ may vary from element to element. In addition, the function $F(x, y)$ can change during contact interaction (for example, in a wear process). Not only do such numerical solutions consume computer time, but they are not universal. A solution for one set of contact characteristics and environment (load, temperature, etc.) cannot be used for another set.

So, it is more important to develop analytical methods to study the discrete contacts under various interface conditions. It allows us to analyze the influence of different microgeometry parameters (density of contact spots, asperities height distribution) on the contact characteristics, friction force, heat conduction, adhesion, wear, and so on, and also to control the friction and wear processes based on the surface engineering.

For these reasons, the discrete contact problem is usually investigated in a simplified formulation. First of all, some model of a real rough surface is considered. The model and the real surface are assumed to be adequate if chosen characteristics of the real surface coincide with the corresponding characteristics of the model.

It is traditional for tribology to model a rough surface as a system of asperities of a regular shape, the space distribution of which reflects the distribution of material in the surface rough layer. Researchers use various shapes of asperities in their models. A complete list of asperity shapes, with their advantages and disadvantages, is given by Kragelsky et al. (1982). The shape of each asperity is determined by a number of parameters (a sphere by its radius, an ellipsoid by the lengths of its axes, etc.). These parameters are calculated from the measurement data of the surface microgeometry. The spacing of the asperities is calculated using the chosen asperity shape and the characteristics of the surface microgeometry obtained from the measurements (Demkin, 1970).

In addition to the approximate description of the surface microgeometry (its roughness), approximate methods of solution of Eqs. (1.1), (1.3), and (1.4) are used to analyze the discrete contact problem. The first studies in the mechanics of discrete contact did not account for the interaction between contact spots; that is, the stress-strain state of bodies in the vicinity of one contact spot was determined by the load applied to this contact, neglecting the deformation caused by the loads applied to the remaining asperities. Under this assumption, the operator A in Eq. (1.1) depends only on the function $p_i(x, y)$, if $(x, y) \in \omega_i$. This assumption gives good agreement between theory and experiment for low contact density, i.e., for low ratio of the real contact area to the nominal one. However, under certain conditions, there are discrepancies between experimental results and predictions. For example, investigating the contact area of elastomers, Bartenev and Lavrentiev (1972) revealed the effect of saturation; that is, the real contact area A_r is always smaller than the nominal contact area A_a, however great a compression load is used. Based on the experimental data, they obtained the following relation:

$$\lambda_A = 1 - \exp\left(-\frac{\beta p}{E}\right), \tag{1.5}$$

where $\lambda_A = A_r/A_a$ is the relative contact area, β is the parameter of roughness, p is the nominal contact pressure, and E is the elasticity modulus of the elastomers. It follows from Eq. (1.5) that $\lambda_A < 1$ for finite values of p. To describe the saturation effect, the mutual influence of the contact spots must be taken into account.

In discrete contact mechanics, the approach developed by Greenwood and Williamson (1966) is widely used. They considered a model of a rough surface consisting of a system of spherical asperities of equal radii, the height of an asperity being a random function with some probability distribution. The deformation of each asperity obeyed the Hertz equation. The additional displacement of the surface because of the average (nominal) pressure distribution within the nominal contact area was also taken into account in this model. So the contact pressures calculated from the model do not satisfy Eqs. (1.2) and (1.3), but the model allows one to estimate approximately the mutual influence effect on the discrete contact characteristics (radius of contact spots, additional displacement, etc.).

1.2 Periodic contact problems

If a surface has clearly defined direction and periodicity of roughness (e.g., at finishing polishing, milling or turning), its microshape can be modeled by regular waviness of various shapes. A surface waviness in general case can be represented by periodic functions. Since for finishing processing the period of waviness is much higher than its amplitude, the linear elasticity can be applied to solve the periodic contact problem. For surfaces with regular microgeometry (for example, wavy surfaces) the methods of solution of periodic contact problems can be used to analyze Eqs.(1.1), (1.3), and (1.4). Periodic contact problems play an important role in studying the effect of mutual influence of contact spots on the characteristics of contact interaction (contact pressure distribution, contact area), and also in analyzing the effect of microgeometry parameters on the approach of the contacting bodies.

1.2.1 2-D periodic contact problems

For a system of rigid punches with flat base penetrating into an elastic half-plane, the 2-D periodic contact problem was first solved by Sadowsky (1928) using a complex stress function. The following expression was derived for the contact pressure:

$$p(x) = \frac{P\left|\cos\left(\frac{\pi x}{L}\right)\right|}{L\sqrt{\sin^2\left(\frac{\pi a}{L}\right) - \sin^2\left(\frac{\pi x}{L}\right)}}, \tag{1.6}$$

where P is the applied normal load per unit length, L is the distance between the punches, and a is the half-width of a punch.

The solution of the periodic contact problem for the inclined punches with the flat base in the case of the complete contact was obtained by Block and Keer (2008). The contact pressure was given by the following expression (Block and Keer, 2008):

$$p(x) = \frac{\pi E\varepsilon\sqrt{2}\sin\left(\frac{\pi x}{L}\right)}{2\pi(1-\nu^2)\sqrt{\cos\left(\frac{2\pi x}{L}\right) - \cos\left(\frac{2\pi a}{L}\right)}} + \frac{P\sqrt{2}\cos\left(\frac{\pi x}{L}\right)}{L\sqrt{\cos\left(\frac{2\pi x}{L}\right) - \cos\left(\frac{2\pi a}{L}\right)}}. \tag{1.7}$$

The normal load P, which acted on each punch and provided the complete contact between the punches inclined at the angle ε and the half-space, satisfied the inequality

$$P \geq \frac{L\varepsilon E}{2(1-\nu^2)}\tan\left(\frac{\pi a}{L}\right) \tag{1.8}$$

In the simplest case the waviness can be described by sinusoid, i.e., $F(x) = \Delta(1 - \cos(2\pi x/l))$, where Δ is an amplitude and l is a period. The contact problem for such a surface, penetrating into an elastic half-plane, was first solved by Westergaard (1939) under the assumption of no friction within the contact spots. The contact pressure $p(x)$ for this type of model roughness is described by the following periodic function (Westergaard, 1939):

$$p(x) = \frac{\sqrt{2}\pi E^* \Delta}{l} \cos\left(\frac{\pi x}{l}\right) \sqrt{\cos\left(\frac{2\pi x}{l}\right) - \cos\left(\frac{2\pi a}{l}\right)} \tag{1.9}$$

Here E^* is the reduced elastic modulus of the contacting bodies, which is determined by the formula:

$$\frac{1}{E^*} = \frac{1 - \nu_1^2}{E_1} + \frac{1 - \nu_2^2}{E_2} \tag{1.10}$$

Here E_i and ν_i are the elastic modulus and the Poisson ratio of the elastic half-space ($i = 1$) and the wavy elastic counterbody ($i = 2$).

The half-width a of the contact zone in Eq. (1.9) is related to the nominal pressure $\overline{p}$ by the following expression (Westergaard, 1939):

$$\overline{p} = \frac{\pi E^* \Delta}{l} \sin^2\left(\frac{\pi a}{l}\right) \tag{1.11}$$

A similar solution for calculating the contact pressure in indentation of a wavy surface described by the periodic function was obtained by Staierman (1949). Later, by using the method of paired summation equation, the normal contact problem for a body with sinusoidal wavy surface and the elastic half-plane was solved (Dundurs et al., 1973), and the following relationship between the maximum contact pressure p_{max} and the applied nominal pressure $\overline{p}$ was derived:

$$p_{max} = 2\sqrt{p^* \overline{p}}, \tag{1.12}$$

where $p^* = \frac{\pi E^* \Delta}{l}$ is the nominal pressure required for the complete contact of the interacting bodies.

The general method for solving a 2-D periodic contact problem for elastic bodies under the assumption of no friction in the contact zone was proposed by Block and Keer (2008). It was based on the reduction of the integral equation of the periodic contact problem, which had the following form (Staierman, 1949):

$$\frac{E^*}{2}g'(x) = \frac{1}{2\pi}\int_{-a}^{a} p(\xi)\cot\left(\frac{x-\xi}{2}\right)d\xi, \tag{1.13}$$

where $g'(x)$ is the derivative of the initial gap function, to the equation corresponding to the problem with a unit contact region:

$$E^* g'(v) = \frac{2}{\pi}\int_{-\alpha}^{\alpha} \frac{p(u)}{v-u}du, \tag{1.14}$$

where $u = \tan\left(\xi/2\right)$, $v = \tan\left(x/2\right)$ and $\alpha = \tan\left(a/2\right)$.

Using this method, the contact problem for the elastic half-plane and the rigid body with the wavy surface described by the function

$$F(x) = \Delta\left[1 - \frac{(m+1)\cos\left(2\pi x/l\right)}{\left|m\cos\left(2\pi x/l\right)\right| + 1}\right], \tag{1.15}$$

where m is the shape parameter $(m < 1)$, was analyzed in (Tsukanov, 2018a). The resulting expression for the contact pressure is a generalization of the Westergaard solution for a sinusoidal wavy surface:

$$p(x) = G(x)\frac{\sqrt{2}\pi\Delta E^*}{l}\left|\cos\left(\frac{\pi x}{l}\right)\right|\sqrt{\cos\left(\frac{2\pi x}{l}\right) - \cos\left(\frac{2\pi a}{l}\right)}, \tag{1.16}$$

where

$$G(x) = (m+1)^2\left(\left|m\cos\left(\frac{2\pi x}{l}\right)\right| + 1\right)^{-2}\left(m\cos\left(\frac{2\pi a}{l}\right) + 1\right)^{-1} \tag{1.17}$$

The effect of the shapes of wavy surface on the contact pressure has been analyzed (Tsukanov, 2018a, 2018b) based on Eq. (1.16) for a rigid body with two-level wavy surface penetrating into an elastic half-plane. In particular, it was shown that the presence of a small-amplitude high-frequency sinusoidal harmonic (second level) on the fundamental one (first level) leads to an oscillating character not only of the contact pressure, but also of the integral contact characteristics (in particular, the dependence of the real contact area on the nominal pressure).

Kuznetsov (1976) was the first to obtain an exact solution of the 2-D periodic contact problem with friction for a rigid wavy indenter and an elastic half-plane. The Kolosov–Muskhelishvili method and the theory of automorphic functions were used to solve the contact problem. For an indenter with a surface profile

$$f(x) = \frac{l^2}{2R\pi^2}\sin^2\left(\frac{\pi}{l}(x-\gamma)\right), \tag{1.18}$$

where l is a period, R is the profile radius at the initial point of contact, and γ is the contact shift, the following expression for the contact pressure within the contact region $(-a-\gamma,\ a-\gamma)$ has been derived (Kuznetsov and Gorohovskii, 1977):

$$p(x) = \frac{4Gl\cos\pi\alpha}{\pi R(\kappa+1)}\cos\left[\frac{\pi}{l}(x+2\alpha a - a\gamma)\right]$$
$$\sin^{1/2-\alpha}\left[\frac{\pi}{l}(a+x)\right]\sin^{1/2+\alpha}\left[\frac{\pi}{l}(a-x)\right]. \tag{1.19}$$

Here a is the half-width of the contact zone, G and ν are the shear modulus and Poisson ratio, respectively, α is the parameter calculated from the relationship

$$\tan(\pi\alpha) = \mu\frac{\varkappa-1}{\varkappa+1}, \tag{1.20}$$

and μ is the friction coefficient. The contact shift γ is calculated from the following expression:

$$\gamma = \frac{l}{2\pi}\arctan\frac{\sin\dfrac{4\pi a\alpha}{l} + 2\alpha\sin\dfrac{2\pi a}{l}}{\cos\dfrac{4\pi a\alpha}{l} + \cos\dfrac{2\pi a}{l}}. \tag{1.21}$$

The combined effect of the friction coefficient and contact density a/l on the contact pressure distribution, as well as on the size and position of the contact regions, was analyzed (Kuznetsov and Gorohovskii, 1977). Based on an approximate solution of this problem (Kuznetsov and Gorokhovsky, 1978), the stress-strain state of the subsurface layers of contacting bodies was studied at different values of the friction coefficient and width of the contact region (Kuznetsov and Gorokhovsky, 1981).

Similar problems for an elastic half-plane in frictional contact with a sinusoidal punch and with a periodic system of punches having a flat base were considered by Block and Keer (2008). The problems were solved by reducing the basic equation to an integral equation of the second kind with the Cauchy kernel. For a sinusoidal punch, they obtained the following expression for the contact pressure at a waviness period $l = 2\pi$:

$$p(x) = \frac{\Delta E^{*}\cos(a/2)\cos(x/2)}{\sin(\pi m)} \sin\left(l + \gamma - \frac{x}{2}\right)\left(\tan\left(\frac{a}{2}\right) - \tan\left(\frac{x}{2}\right)\right)^{m}\left(\tan\left(\frac{a}{2}\right) + \tan\left(\frac{x}{2}\right)\right)^{1-m} \quad (1.22)$$

Here the eccentricity γ of the contact region is calculated from the expression

$$\gamma = \tan^{-1}\left(\frac{\cos(l)}{\sin(l) - \psi^{2}\sqrt{1 + \psi^{2}m(1 - m)}}\right) \quad (1.23)$$

In Eqs. (1.22) and (1.23), $\tan \pi m = 1/\beta\mu$, $l = (\pi m - \phi(2m - 1))$, and $\tan \phi = 1/\psi$.

For a periodic system of punches with a flat base (a is the half-width of the punch base), the contact pressure distribution is obtained in the form (Block and Keer, 2008):

$$p(x) = \frac{P\sin\pi m\left(\tan\left(\frac{a}{2}\right) - \tan\left(\frac{x}{2}\right)\right)^{m-1}\left(\tan\left(\frac{a}{2}\right) - \tan\left(\frac{x}{2}\right)\right)^{-m}}{2\pi\cos\left(\frac{a}{2}\right)\sin(\pi m - \phi(2m - 1))}. \quad (1.24)$$

A general formulation and a method for solving various contact problems, including periodic ones, considering adhesion and slippage by reducing them to a vector Riemann problem, were proposed by Antipov and Arutyunyan (1991). In this and their earlier studies, it was shown that the mathematical formulation of the problems taking into account friction in contact regions can be reduced to a coupled system of integral equations for the normal and shear contact stresses. In a special case of equal elastic constants of the contacting materials, the system is uncoupled.

1.2.2 3-D periodic contact problems

3-D periodic contact problems are of considerable interest because most rough surfaces in nature and technical applications are isotropic; i.e., the height and step parameters of the roughness profiles in the longitudinal and transverse directions are comparable in magnitude. In addition, 3-D periodic contact problems arise when there is a special texture on the contacting surfaces, as a rule, formed by physical and chemical treatment, e.g., by laser or pressure. From the viewpoint of elasticity theory, the 3-D periodic contact problem is much more complicated than the 2-D one due to the absence of a direct resolvent of the basic integral equation followed from Eqs. (1.2) and (1.3). Often Eqs. (1.1) and (1.3) are solved using iterative procedures or the boundary element method. With complete contact between the bodies, the problem becomes linear and much simpler.

Johnson et al. (1985) developed a method of analysis of a discrete contact problem for an elastic body, the surface of which in two mutually perpendicular directions was described by two sinusoidal functions; the counter body had a smooth surface. Based on the superposition principle, it was shown that, for a regular relief described by the function

$$f(x, y) = \Delta_1 + \Delta_2 - \Delta_1 \cos\frac{2\pi x}{l_1} - \Delta_2 \cos\frac{2\pi y}{l_2}, \tag{1.25}$$

the contact pressure at complete contact is expressed as

$$p(x, y) = \overline{p} + p_x^* \cos\frac{2\pi x}{l_1} + p_y^* \cos\frac{2\pi y}{l_2}, \tag{1.26}$$

where $p_x^* = \pi E^* \Delta_1 / l_1$ and $p_y^* = \pi E^* \Delta_2 / l_2$, and E^* is the equivalent elastic modulus determined by Eq. (1.10). To provide the complete contact, the following condition for the nominal pressure must be satisfied: $\overline{p} \geq p_x^* + p_y^*$. For lower values of the nominal pressure $\overline{p}$, incomplete contact occurs. Asymptotic methods to solve the contact problem were developed for low values of $\overline{p}$ based on the Hertz theory and for values close to the limited value $p_x^* + p_y^*$ (but not exceeding this limit) based on the elliptic cracks theory (Johnson et al., 1985). For intermediate values of $\overline{p}$ the solution was constructed numerically based on seeking the Fourier coefficients with minimizing the total additional energy (similar to the case of the 2-D problem for sinusoidal waviness).

Using the finite element method, relationships for integral contact characteristics, which are close to the results of Johnson et al. (1985), were obtained in (Rostami and Jackson, 2013); in addition, expressions that approximate the real contact area and mean gap between the corrugated surface and the half-space for the entire range of applied nominal pressures were proposed. The results of Johnson et al. (1985) on the contact problem for a wavy surface (1.25) were refined in (Yastrebov et al., 2014) by using the boundary element method and the fast Fourier transform. Inflection points were found on the real contact area as a function of the nominal pressure. The first point corresponds to the transformation of the contact region's contour from a circle to a square with rounded corners. The second one corresponds to the merging of adjacent contact regions.

Based on the expansion of function (1.25) in a double Fourier series in polar coordinates, the asymptotic dependences obtained by Johnson et al. (1985) were refined by Tsukanov (2019). The obtained relationships for the real contact area and the mean gap as functions of the nominal pressure correlate well with the results of numerical simulation (Rostami and Jackson, 2013; Yastrebov et al., 2014).

1.3 Approximate solution of the periodic contact problem based on the method of localization

In this section, an approach to derive an approximate solution of the discrete contact problem for periodic systems of indenters based on the localization method is presented.

1.3.1 Problem formulation for one-level periodic system of axisymmetric smooth indenters

A periodic contact problem for a system of axisymmetric smooth indenters, the contacting surface of which is described by an axisymmetric function $z = f(r)$, that interacts without friction with an elastic half-space is considered (Fig. 1.2).

The axes of the indenters are perpendicular to the half-space surface $z = 0$ and intersect this surface at points that are distributed uniformly over the plane $z = 0$. As an example of such a system, we can consider indenters located at the sites of a quadratic or hexagonal lattice.

Let us fix an arbitrary indenter and locate the origin O of a polar system of coordinates (r, θ) in the plane $z = 0$ at the point of intersection of the axis of this indenter with the plane $z = 0$ (see Fig. 1.2). The tops of the indenters have the coordinates (r_i, θ_{ij}) $(i = 1, 2, \dots\ j = 1, 2, \dots, m_i)$ where m_i is the number of indenters located at the circumference of the radius r_i $(r_i < r_{i+1})$.

Due to the periodicity of the problem, each contact occurs under the same conditions. We assume that contact spots are circles of a radius a

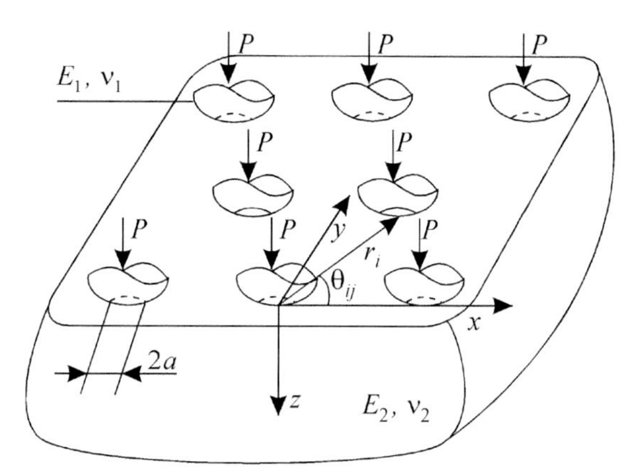

Figure 1.2 Scheme of the periodic contact between smooth indenters and a half-space.

(it imposes certain restrictions on the density of the indenters arrangement), and that only normal pressure $p(r,\theta)$ acts at each contact spot $(r \le a)$ (the shear stress is negligibly small). To determine the pressure $p(r,\theta)$ acting at an arbitrary contact spot with a center O, we use the solution of a contact problem for an axisymmetric indenter with surface shape $z = f(r)$ and an elastic half-space subjected to the pressure $q(r,\theta)$, distributed outside the contact region, which was derived by Galin (1953, 2008). The contact pressure $p(r,\theta)$ $(r \le a)$ is determined by the following relationship (Galin, 1953):

$$p(r,\theta) = G(r) + \frac{c(\theta)}{\sqrt{a^2 - r^2}} - \frac{1}{\sqrt{a^2 - r^2}} \int_a^{+\infty} \int_0^{2\pi} q(r',\theta') H_2(r,\theta,r',\theta') r' dr' d\theta', \tag{1.27}$$

where

$$G(r) = \frac{E^*}{4\pi^2} \int_0^a \Delta f(r') H_1(r,r') dr', \tag{1.28}$$

$$H_1(r,r') = \int_0^{2\pi} \frac{2r'}{\sqrt{r^2 - 2rr'\cos\theta' + r'^2}} \arctan \frac{\sqrt{a^2 - r^2}\sqrt{a^2 - r'^2}}{a\sqrt{r^2 - 2rr'\cos\theta' + r'^2}} d\theta', \tag{1.29}$$

$$H_2(r,\theta,r',\theta') = \frac{\sqrt{r'^2 - a^2}}{\pi^2[r^2 + r'^2 - 2rr'\cos((\theta-\theta')]}, \tag{1.30}$$

and E^* is determined by Eq. (1.10). The function $c(\theta)$ depends on the shape of the indenter $f(r)$. For example, if the indenter is smooth (the function $f'(r)$ is continuous at $r \le a$), then the contact pressure is zero at $r = a$, i.e., $p(a,\theta) = 0$, and the function $c(\theta)$ has the form

$$c(\theta) = \int_a^{+\infty}\int_0^{2\pi} q(r',\theta')H_2(a,\theta,r',\theta')r'dr'd\theta'. \tag{1.31}$$

The first term in Eq. (1.27) means the pressure that acts under a single axisymmetric indenter of the shape function $f(r)$ penetrating into an elastic half-space, and the last two terms are the additional contact pressure occurring due to the pressure $q(r,\theta)$ distributed outside the contact region.

For the periodic contact problem, the function $q(r,\theta)$ coincides with the pressure $p(r,\theta)$ at each contact spot located at $(r_i,\theta_{ij})(r_i > a)$, and is zero outside contact spots. So we obtain the following integral equation from Eq. (1.27), under the assumption that $f'(r)$ is a continuous function $(p(a,\theta) = 0)$:

$$p(r,\theta) - \int_0^a\int_0^{2\pi} K(r,\theta,r',\theta')p(r',\theta')r'dr'd\theta' = G(r), \tag{1.32}$$

where

$$K(r,\theta,r',\theta') = \sum_{i=1}^{\infty} K_i(r,\theta,r',\theta'), \tag{1.33}$$

and

$$K_i(r,\theta,r',\theta') = \frac{1}{\pi^2\sqrt{a^2-r^2}}\sum_{j=1}^{m_i}\left[K_{ij}(a,\theta,r',\theta') - K_{ij}(r,\theta,r',\theta')\right], \tag{1.34}$$

$$K_{ij}(r,\theta,r',\theta') = \frac{\sqrt{r_i^2 + r'^2 + 2r_ir'\cos(\theta_{ij}-\theta') - a^2}}{(r\cos\theta - r'\cos\theta' - r_i\cos\theta_{ij})^2 + (r\sin\theta - r'\sin\theta' - r_i\sin\theta_{ij})^2}. \tag{1.35}$$

The kernel $K(r,\theta,r',\theta')$ in Eq. (1.32) is represented as a series (1.33). A general term (1.34) of this series can be transformed to the form:

$$K_i(r,\theta,r',\theta') = \frac{1}{\pi^2\sqrt{a^2-r^2}}\sum_{j=1}^{m_i}\left\{\frac{2(a-r)\cos\left(\theta_{ij}-\theta\right)}{r_i^2}\right.$$
$$\left.+\frac{(a-r)\left[-a-r-6r'\cos\left(\theta_{ij}-\theta'\right)\cos\left(\theta_{ij}-\theta\right)+2r'\cos(\theta'-\theta)\right]}{r_i^3}+O\left(\frac{1}{r_i^4}\right)\right\}. \tag{1.36}$$

We assume that for the periodic system of indenters under consideration, each contact spot with the center $\left(r_i;\theta_{ij}\right)$ has a partner with the center at the point $\left(r_i;\pi+\theta_{ij}\right)$. So the sum on the first line of Eq. (1.36) is zero. Hence, the general term of the series in (1.33) has the order $O\left({}^{1}/{}_{r_i^2}\right)$, since $m_i\sim r_i$, and the series converges.

1.3.2 Method of localization

Let us consider the following equation:

$$p(r,\theta)-\int_0^a\int_0^{2\pi}\sum_{i=1}^{n}K_i(r,\theta,r',\theta')p(r',\theta')r'dr'd\theta'$$
$$=G(r)+\frac{2\overline{N}P}{\pi}\arctan\frac{\sqrt{a^2-r^2}}{\sqrt{A_n^2-a^2}}, \tag{1.37}$$

where $\overline{N}$ is the average number of contact spots per unit area, and P is a load applied to each contact spot. This load satisfies the equilibrium equation

$$P=\int_0^a\int_0^{2\pi}p(r,\theta)rdrd\theta. \tag{1.38}$$

To obtain Eq. (1.37), we replace the summation over the region Ω_n ($\Omega_n:\ r\geq A_n,\ 0\leq\theta\leq 2\pi$) by integration over $i>n$ in Eq. (1.33), taking into account that the centers of contact spots are distributed uniformly over the plane $z=0$ and their number per unit area is characterized by the value $\overline{N}$. The following transformation demonstrates how Eq. (1.37) is derived:

$$J_n = \sum_{i=n+1}^{\infty} K_i(r,\theta,r',\theta') \approx \overline{N} \int_{A_n}^{+\infty} \int_0^{2\pi} \frac{\sqrt{x^2 + r'^2 + 2xr'\cos(\phi - \theta') - a^2}}{\pi^2\sqrt{a^2 - r^2}}$$

$$\times \left[\frac{1}{(a\cos\theta - r'\cos\theta' - x\cos\phi)^2 + (a\sin\theta - r'\sin\theta' - x\sin\phi)^2} - \frac{1}{(r\cos\theta - r'\cos\theta' - x\cos\phi)^2 + (r\sin\theta - r'\sin\theta' - x\sin\phi)^2} \right] x dx d\phi. \tag{1.39}$$

Changing the variables $y\cos\psi = x\cos\phi + r'\cos\theta'$, $y\sin\psi = x\sin\phi + r'\sin\theta'$, and taking into account that $r' \leq a \ll A_n$, we finally obtain

$$J_n \approx \frac{\overline{N}}{\pi^2\sqrt{a^2 - r^2}} \int_{A_n}^{+\infty} \int_0^{2\pi} \sqrt{y^2 - a^2} \left[\frac{1}{a^2 + y^2 - 2ay\cos\psi} - \frac{1}{r^2 + y^2 - 2ry\cos\psi} \right] y dy d\psi = \frac{2\overline{N}}{\pi} \arctan \frac{\sqrt{a^2 - r^2}}{\sqrt{A_n^2 - a^2}}, \tag{1.40}$$

where A_n is the radius of a circle in which there are $m_1 + m_2 + ... + m_n + 1$ indenters. It is apparent that

$$A_n^2 = \frac{1}{\pi \overline{N}} \left(\sum_{i=1}^{n} m_i + 1 \right). \tag{1.41}$$

We note that the solution of Eq. (1.37) tends to the solution of Eq. (1.32) as $n \to \infty$.

Let us analyze the structure of Eq. (1.37). The integral term on the left side of Eq. (1.37) governs the influence of the real pressure distribution over the neighboring contact spots $(r_i < A_n)$ on the pressure at a fixed contact spot with the center $(0,0)$ (local effect). The effect of the pressure distribution over the remaining contact spots with centers (r_i, θ_{ij}), $r_i > A_n$ is taken into account by the second term in the right side of Eq. (1.37). This term describes the additional pressure $p_a(r)$ that arises within a contact spot $(r < a)$ from the nominal pressure $\overline{p} = P\overline{N}$ in the region Ω_n $(r > A_n)$. Indeed, from Eqs. (1.27) and (1.31), it follows that the additional pressure $p_a(r)$ within the contact spot $(r \leq a)$ arising from the pressure $q(r,\theta) = \overline{p}$ distributed uniformly in the region Ω_n has the form

$$p_a(r) = \frac{\overline{p}}{\pi^2\sqrt{a^2 - r^2}} \int_{A_n}^{+\infty} \int_0^{2\pi} \sqrt{r'^2 - a^2}\left[\frac{1}{a^2 + r'^2 - 2ar'cos\theta} - \frac{1}{r^2 + r'^2 - 2rr'cos\theta}\right] r'dr'd\theta = \frac{2\overline{p}}{\pi}\arctan\frac{\sqrt{a^2 - r^2}}{\sqrt{A_n^2 - a^2}}. \quad (1.42)$$

Thus, the effect of the real contact pressure distribution over the contact spots ω_i far away from the contact spot under consideration can be taken into account to sufficient accuracy by the nominal pressure $\overline{p}$ distributed over the region Ω_n. It is illustrated in Fig. 1.3.

This conclusion drawn for the periodic contact problem under consideration is a particular case of a more general statement that we call the method of localization (Goryacheva, 1998): in the conditions of discrete contact, the stress-strain state near any fixed contact spot can be calculated to sufficient accuracy by taking into account the real contact conditions (real pressure, shape of bodies, etc.) at this contact spot and at the nearby contact spots (in the local vicinity of the fixed contact), and the averaged (nominal) pressure over the remaining part of the region of interaction (nominal contact region). As it will be shown in this chapter, this method is

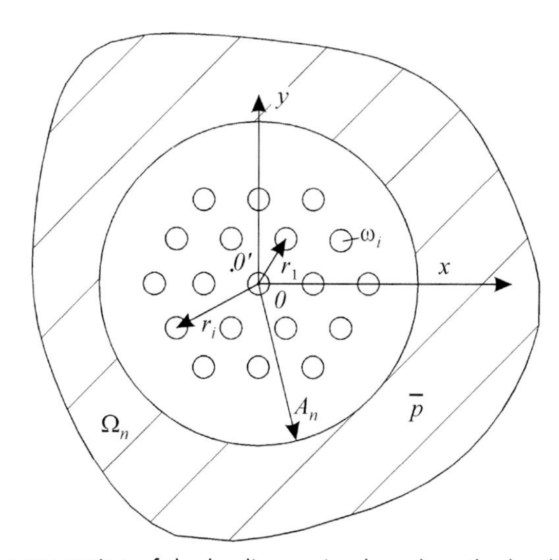

Figure 1.3 Representation of the loading region based on the localization method.

applicable also for solving discrete contact problems with bounded nominal region.

Eqs. (1.37) and (1.38) are used to determine the contact pressure $p(r,\theta)$ and the radius a of each contact spot. The stress distribution in the sub-surface region $(z>0)$ arising from the real contact pressure distribution at the surface $z=0$ can then be found by superposition, using the potentials of Boussinesq (1885) or the particular solution of the axisymmetric problem given by Timoshenko and Goodier (1951).

To simplify the procedure, we can use the principle of localization for calculating the internal stresses, by replacing the real contact pressure at distant contact spots with the nominal contact pressure. We give here the analytical expressions for the additional stresses that occur on the axis of symmetry of any fixed contact spot from the action of the nominal pressure $\bar{p}$ within the region Ω_n $(r > A_n)$:

$$\sigma_z = -\frac{\bar{p}z^3}{\left(A_n^2+z^2\right)^{3/2}},\ \sigma_r=\sigma_\theta=\frac{\bar{p}z}{\sqrt{A_n^2+z^2}}\left[\frac{z^2}{2\left(A_n^2+z^2\right)}-(1+\nu)\right],$$
$$\tau_{rz}=\tau_{\theta z}=\tau_{r\theta}=0. \tag{1.43}$$

1.3.3 Analysis of the contact and internal stresses (one-level periodic system of indenters)

We use the relationships obtained above to analyze the real contact pressure distribution and internal stresses for the periodic contact problem for a system of indenters and the elastic half-space. Particular emphasis will be placed upon the influence of the geometric parameter, which describes the density of indenter location, on the contact characteristics. This will allow us to determine the range of parameter variations in which it is possible to use the simplified theories that neglect the interaction between contact spots (the integral term in Eq. (1.32)) or the local effect of the influence of the real pressure distribution within neighboring contact spots on the pressure at the fixed spot (the integral term in Eq. (1.37)).

Numerical results are presented here for a system of spherical indenters, ($f(r)=r^2/2R$, R is a radius of curvature) located at sites of a hexagonal lattice with a constant pitch l. In this case the density of indenter arrangement is $\overline{N}=2/\left(l^2\sqrt{3}\right)$.

We introduce the following dimensionless parameters and functions

$$\rho=\frac{r}{R},\ A_n^1=\frac{A_n}{R},\ a^1=\frac{a}{R},\ l^1=\frac{l}{R},\ p^1(\rho,\theta)=\frac{\pi p(\rho R,\theta)}{2E^*},\ P^1=\frac{\pi P}{2E^*R^2}. \tag{1.44}$$

The system of Eqs. (1.37) and (1.38) was solved by iteration.

To determine the radius A_n of the circle ($r \le A_n$), where the real pressure distribution within nearby contact spots is taken into account (local effect), and the corresponding value of n that gives an appropriate accuracy of the solution of Eq. (1.37), we calculated the contact pressure $p^1(\rho,\theta)$ from Eqs. (1.37) and (1.38) for $n = 0$, $n = 1$, $n = 2$, and so on. For $n = 0$, the integral term on the left-hand side of Eq. (1.37) is zero, so that the effect of the remaining contact spots surrounding the fixed one (with the center at the origin of coordinate system O) is taken into account by a nominal pressure distributed outside the circle of radius A_0 (the second term in the right side of Eq. (1.37), where A_0 is specified by Eq. (1.41)). In this case Eq. (1.37) takes the form:

$$p(r)=\frac{E^*}{4\pi^2}\int_0^a \Delta f(\rho)H_1(r,\rho)d\rho+\frac{2}{\pi}\overline{N}P\arctan\left(\frac{\sqrt{a^2-r^2}}{\sqrt{A^2-a^2}}\right) \tag{1.45}$$

For $n = 1$ we take into account the real pressure within six contact spots located at the distance l from the fixed one; for $n = 2$, there are 12 contact spots, six located at the distance l and another six at a distance $l\sqrt{3}$, and so on. Fig. 1.4 illustrates the results calculated for $a^1 = 0.1$ and $l^1 = 0.2$, i.e., $a/l = 0.5$, and this case corresponds to the limiting value of contact density. The results show that the contact pressures calculated for $n = 1$ and $n = 2$ differ from one another by less than 0.1%. If the contact density decreases (a/l decreases), this difference also decreases. Based on this estimation, we will take $n = 1$ in the further analysis.

First, we analyze the effect of interaction between contact spots on the pressure distribution. Fig. 1.5 illustrates the contact pressure under an indenter of one-level system for different values of the parameter l^1 characterizing the spacing between indenters. In all cases, the normal load $P^1 = 0.0044$ is applied to each indenter.

The results show that the radius of the contact spot decreases and the maximum contact pressure increases if the spacing l between indenters decreases; the contact density characterized by the parameter a/l also increases: $a/l = 0.128$ (curve 3), $a/l = 0.45$ (curve 2), and $a/l = 0.5$ (curve 1). Curve 3 almost coincides with the contact pressure distribution calculated from the

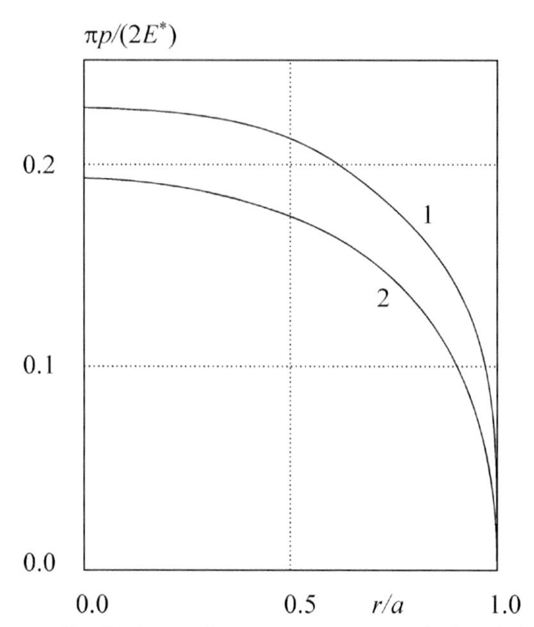

Figure 1.4 Pressure distribution within a contact spot, calculated from Eq. (1.37) for $n = 0$ (curve 1), $n = 1$ and $n = 2$ (curve 2), and $a/R = 0.1$, $l/R = 0.2$ (one-level model).

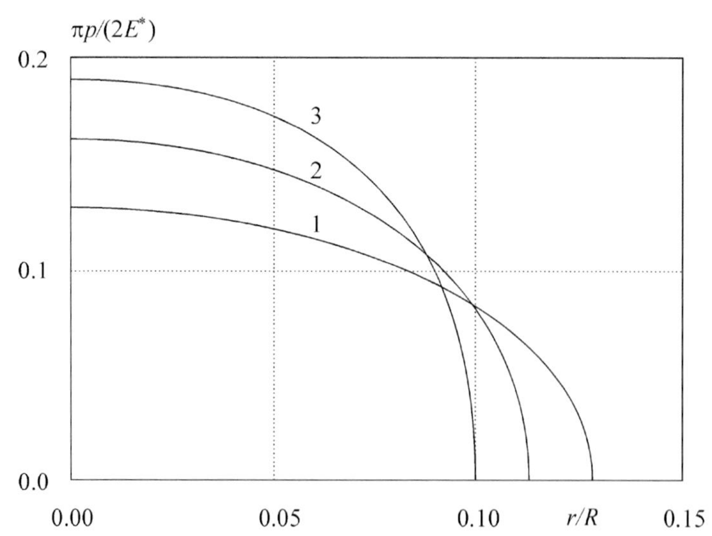

Figure 1.5 Pressure distribution under an indenter acted on by the force $P^1 = 0.0044$ for the one-level model characterized by various spacings between indenters: $l/R = 0.2$ (curve 1), $l/R = 0.25$ (curve 2), and $l/R = 1$ (curve 3).

Hertz theory, which neglects the influence of contact spots surrounding the fixed one. So, for small values of the parameter a/l, it is possible to neglect the interaction between contact spots for calculating the contact pressure.

The dependencies of the radius of a contact spot on the dimensionless nominal pressure $\bar{p}^1 = \bar{p}\pi/(2E^*)$ calculated for different values of the parameter l^1 for a one-level model are shown in Fig. 1.6 (curves 1, 2, 3). The results of calculation based on the Hertz theory are added for comparison (curves 1', 2', 3'). The results show that under a constant nominal pressure $\bar{p}$ the radius of each contact spot and, hence, the real contact area, decreases if the relative distance l/R between contact spots decreases. The comparison of these results with the curves calculated from the Hertz theory makes it possible to conclude that for $\frac{a}{l} < 0.25$ the discrepancy between the results predicted from the discrete contact theory and Hertz theory does not exceed 2.5%. For higher nominal pressure and, hence, higher contact density, the discrepancy becomes significant. Thus, for $l = 0.5$ (curves 2, 2′) and $a/l = 0.44$ the calculation of the real contact area from the Hertz theory gives an error of about 15%.

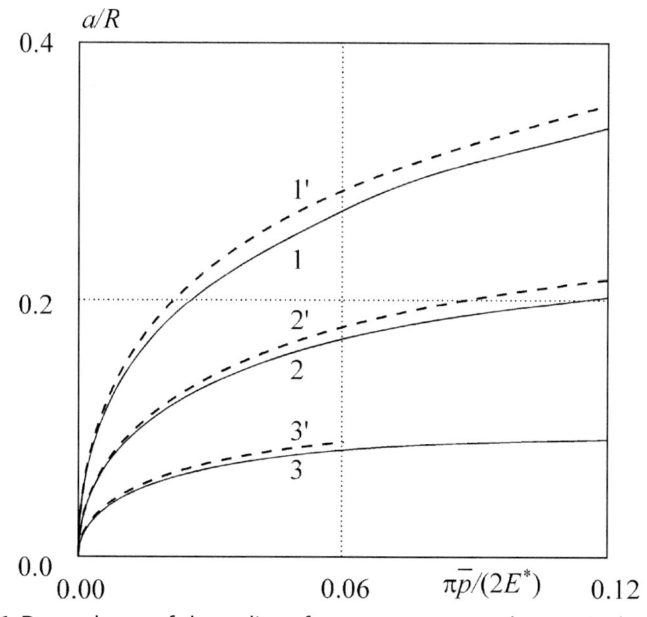

Figure 1.6 Dependence of the radius of a contact spot on the nominal pressure for $l/R = 1$ (curves 1, 1′), $l/R = 0.5$ (curves 2, 2′), and $l/R = 0.2$ (curves 3, 3′), calculated from Eq. (1.37) (1, 2, 3) and from Hertz theory (1′, 2′, 3′).

It follows from the analysis of the internal stresses within the elastic half-space subjected to loading by the periodic system of indenters, that there is a nonuniform stress field in the subsurface layer, the thickness of which is comparable with the spacing l between indenters. The stress field features depend essentially on the contact density parameter a/l. Fig. 1.7 illustrates the dimensionless principal shear stress $\tau_1/\overline{p}$ along the z-axis that coincides with the axis of symmetry of the indenter (curves 1, 2) and along the axis $O'z$ (curves 1', 2') equally spaced from the centers of the contact spots (see Fig. 1.2). The results are calculated for the same nominal pressure $\overline{p}$ and different spacings l/R between indenters: curves 1 and 1' correspond to $l/R = 1$ (or $a/l = 0.35$), and curves 2 and 2' correspond to $l/R = 0.5$ (or $a/l = 0.42$). The results indicate that as the parameter a/l increases (the relative real contact area increases), the maximum value of the principal shear stress near the surface at the fixed depth decreases, and the maximum difference of the principal shear stress at the fixed depth also decreases. At infinity the principal shear stresses depend only on the nominal pressure $\overline{p}$. So the internal stresses differ noticeably from ones calculated from the Hertz model if the parameter a/l varies in the limits $0.25 < a/l < 0.5$. An increase

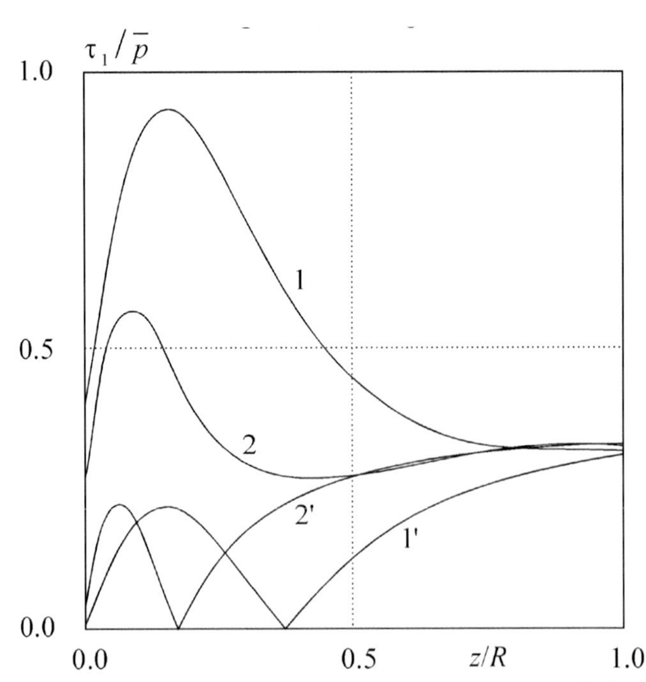

Figure 1.7 The principal shear stress along the axes Oz (curves 1, 2) and $O'z$ (curves 1', 2').

of the density of contact spots leads to appearance of the stressed layer at some depth under the surface. The same conclusion was made by Kuznetsov and Gorohovsky (1978) based on the analysis of internal stresses in the 2-D contact problem for a sinusoidal punch and an elastic half-plane.

The method of localization was also used (Goryacheva and Torskaya, 2003) to study the contact of a periodic system of spherical indenters and a two-layered elastic half-space. The additional geometrical parameter that describes the ratio of the surface layer (coating) thickness H to the spacing l between the contact spots has been included in the analysis. It was shown that this parameter influences significantly the contact and internal stress distributions, particularly for relatively hard and thin coatings.

1.3.4 Three-level periodic system of axisymmetric smooth indenters

The method of localization is also applicable for calculating the real pressure distribution in contact interaction of a periodic system of indenters of various heights and an elastic half-space.

Let us assume that the shape of indenters in a system of coordinates related to the underformed surface of the elastic half-space (Fig. 1.2) is described by the continuously differentiable function $z = f_m(r) + h_m$ where $f_m(r)$ and h_m ($h_1 < h_2 < h_3$) are the shape and the height of indenters of the level m ($m = 1, 2, 3$). Indenters of each level are located at the hexagonal lattice sites with the constant pitch l (see Fig.1.8a).

When only the highest indenters are in the contact, the procedure of the contact characteristics calculation is similar to the case considered in Section 1.3.3 for the one-level periodic system of indenters. With calculating the contact pressure under the highest indenters, the displacement w_1 of the boundary of the elastic half-space at the center of the triangle with the side l is also calculated. When the condition $D - w_1 = h_1 - h_2$ is satisfied (D is the system penetration taking into account the mutual influence of the contact spots of the first level), the second level of indenters comes into contact and so on.

If all the indenters are in contact, we choose any indenter of the m-th level and place the origin of the polar system of coordinates at the center of its contact spot (see Fig. 1.8b). Then we can use the principle of localization and take into account the real pressure $p_j(r, \theta)$ ($j = 1, 2, ..., k$) at the contact spots inside the region $\overline{\Omega}_m$ which is a circle of the radius A_m ($\overline{\Omega}_m$: $r \leq A_m$).

Replacing the real contact pressure distribution within the contact spots at the region $r > A_m$ by the nominal pressure $\overline{p}$,

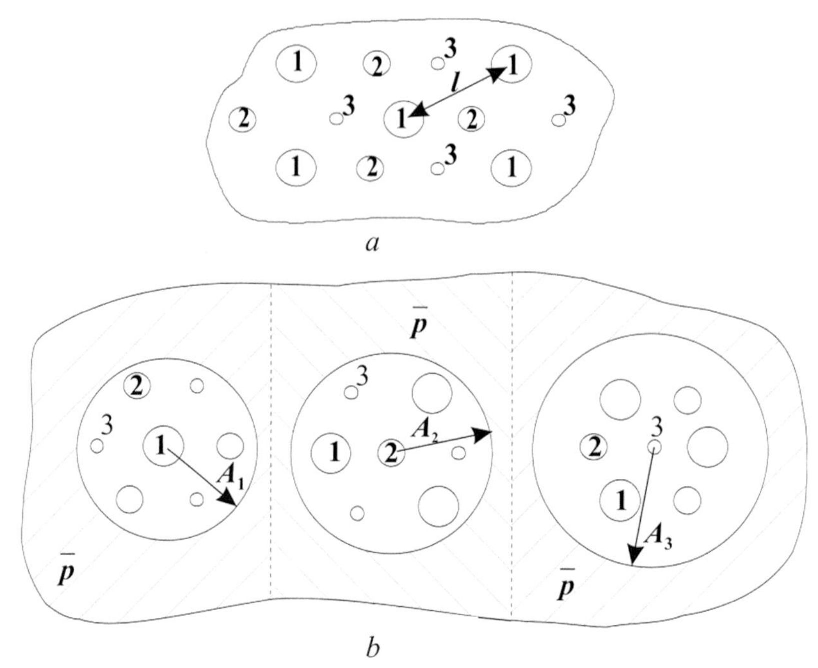

Figure 1.8 Location of indenters of each level ($k = 3$) in the model (a) and scheme of calculations of the pressures acting on the indenters (b).

$$\overline{p} = \sum_{j=1}^{k} \overline{N}_j \int_0^{a_j} \int_0^{2\pi} p_j(r', \theta') r' dr' d\theta',$$

where $\overline{N}_j$ is the number of indenters at the j-th level at the unit area (density of the indenters of the j-th level),
we obtain the relationship that is similar to Eq. (1.37):

$$p_m(r, \theta) - \sum_{j=1}^{k} \int_0^{a_j} \int_0^{2\pi} K_n(a_m, r, \theta, r', \theta') p_j(r', \theta') r' dr' d\theta'$$

$$= G_m(r) + \frac{2\overline{p}}{\pi} \arctan \frac{\sqrt{a_m^2 - r^2}}{\sqrt{A_m^2 - a_m^2}}. \tag{1.46}$$

The kernel of Eq. (1.46) has the form

$$K_n(a_m, r, \theta, r', \theta') = \sum_{i=1}^{n} K_i(a_m, r, \theta, r', \theta'),$$

where functions $K_i\left(a_{m,} r, \theta, r', \theta'\right)$ are determined by Eqs. (1.34) and (1.35) (in this equation, instead of a and $f(r)$, we must put a_m and $f_m(r)$, respectively).

This procedure must be repeated for indenters of each level (see Fig. 1.8b). So we obtain the system of k integral Eq. (1.46) for calculation of the pressures $p_m(r, \theta)$ within the contact spots ($r \leq a_m$) at each m-th level ($m = 1, 2, ...k$). Since the radius a_m of contact spot of each level is also an unknown value, the additional condition taking into account the difference of the heights of indenters of each level is used for the analysis (Goryacheva, 1998). The system of equations is completed by the equilibrium condition.

The calculations were performed for the three-level model of spherical indenters located at the sites of the hexagonal lattice (Fig. 1.8) with prescribed height distribution: $(h_1 - h_2)/R = 0.014$ and $(h_1 - h_3)/R = 0.037$. Fig. 1.9 illustrates the pressure distributions within the contact spots of each level (curves 1, 2, and 3) for the constant value of the dimensionless total load $P^1 = P_1^1 + P_2^1 + P_3^1$ applied to three indenters. The curves $1'$, $2'$, and $3'$ in Fig. 1.8 correspond to the Hertz solution neglecting the mutual influence of contact spots.

The results indicate that the smaller is the height of the indenter, the greater is the difference between the contact pressure calculated from the periodic contact problem and Hertz theory. It is explained by the deflection of the elastic half-space surface between contact spots due to their mutual influence and by strong dependence of this deflection on the contact density parameter. So the load that must be applied to bring a new level of indenters into contact depends not only on the height difference of the indenters, but also on the contact density.

The approach developed was used (Goryacheva and Torskaya, 2019) to study the contact characteristics in interaction of a multilevel periodic system of indenters and coated elastic half-space. Contact pressure distributions under the indenters of different levels, internal stresses, and additional displacements as a function of average pressure have been determined for different values of input parameters: nominal pressure, space distribution of indenters, their contact density, and coating thickness. The mutual effect was analyzed for the case of relatively hard and soft coatings. It was

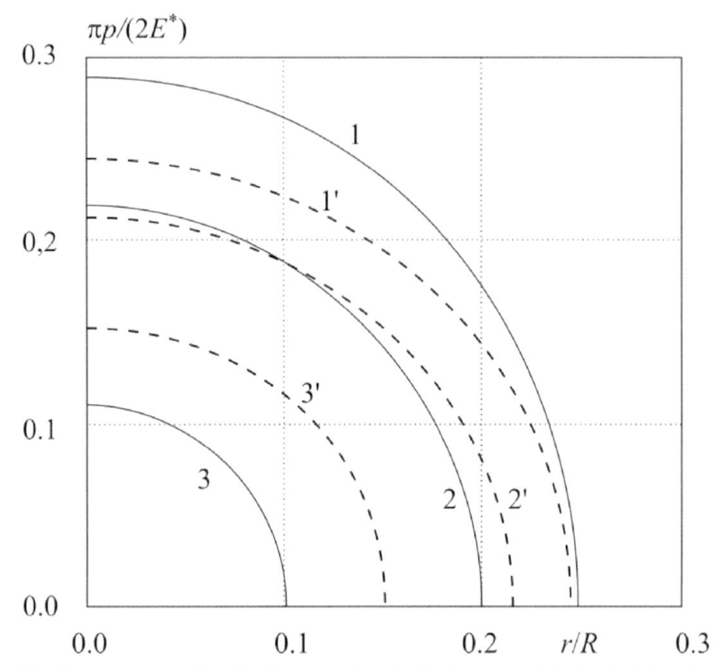

Figure 1.9 The pressure distributions under the indenters with the heights h_1 (curves 1 and 1′), h_2 (curves 2 and 2′), and h_3 (curves 3 and 3′) for the three-level model ($(h_1 - h_2)/R = 0.014$, $(h_1 - h_3)/R = 0.037$, $P^1 = 0.059$) calculated from Eq. (1.46) (curves 1, 2, 3) and from the Hertz theory (curves 1′, 2′, 3′).

concluded from the calculated results that for the case of thin, soft coatings, this effect is stronger for materials with relatively high value of Poisson ratio. For relatively hard, thin coatings, due to the coating bending, the real contact pressure distribution and real contact area strongly depend on the contact density and height distribution of indenters. In this problem, the periodical system of indenters can be considered as a model of the regular distributed asperities of a rough surface.

1.3.5 Application of the method of localization for solving 2-D periodic contact problems

In the general case, the 2-D periodic contact problem without friction is reduced to solving the following integral equation with the Hilbert kernel (Staierman, 1949; Barber, 2018):

$$\frac{E^*}{2} h'(x) = \frac{1}{2\pi} \int_{-a}^{a} p(\xi) \cot \frac{x - \xi}{2} \, d\xi, \tag{1.47}$$

where $h'(x)$ is the derivative of the initial gap function between surfaces, $p(x)$ is the unknown contact pressure distribution, and a is the half-width of the contact zone. The closed form solution of this equation exists only for certain initial gap functions $h(x)$.

To obtain the approximate solution of the periodic contact problem in 2-D formulation, we can also apply the localization principle. In the simplest formulation, it involves the contact problem solution for a single asperity (determination of contact pressure distribution) taking into account the normal displacements of a half-plane boundary inside the considered single contact zone from action of the averaged pressure, applied outside the certain region (see Section 1.3.2).

The contact problem for a single contact segment is reduced to the integral equation with the Cauchy kernel (Muskhelishvili, 1953):

$$\frac{E^*}{2}h'(x) = \frac{1}{2\pi}\int_{-a}^{a}\frac{p(\xi)}{x-\xi}d\xi. \tag{1.48}$$

The solution of this equation for a symmetric smooth initial gap function $h(x)$ and bounded pressure on both ends of the contact segment has the following form (Muskhelishvili, 1953):

$$p(x) = \frac{E^*}{2\pi}\sqrt{a^2 - x^2}\int_{-a}^{a}h'(\xi)\frac{1}{\sqrt{a^2-\xi^2}}\frac{1}{\xi - x}d\xi. \tag{1.49}$$

The initial gap function within the contact zone is determined by the following condition:

$$h(x) = \delta - (f(x) + f_2(x)), \tag{1.50}$$

where $f(x)$ is the asperity shape function, δ is the contact approach, and $f_2(x)$ is the function that describes curvature of the half-plane boundary caused by the action of the remaining asperities except for the one under consideration.

According to the localization principle, the action of the remaining asperities is replaced by a uniform pressure acting outside a strip of width $2b$. The value of $2b$ is determined from the condition of equality of the mean pressure inside and outside of this strip. The mean pressure in a plane periodic contact problem is determined as $\overline{p} = P_s/L$, where P_s is the total load on a single contact segment, and L is the distance between the peaks of asperities (period): therefore, $2b = L$. To determine the total load on a single contact zone P_s, the equilibrium equation is used:

$$P_s = \int_{-a}^{a} p(x)dx. \tag{1.51}$$

The function $f_2(x)$ can be represented as a difference between displacements from a uniform load distributed over the entire half-plane and displacements from the same load inside a strip of width L (Johnson, 1985):

$$f_2(x) = -\frac{2}{\pi E^*}\frac{P_s}{L}\left(C - \int_{-a}^{x} \ln\left[\frac{L/2+\xi}{L/2-\xi}\right]d\xi\right). \tag{1.52}$$

After differentiating Eq. (1.52) and substitution the result into (1.50), we obtain the expression for the derivative of the gap function inside the single contact zone

$$h'(x) = f'(x) + \frac{2}{\pi E^*}\frac{P_s}{L}\left(\ln\left[\frac{L/2+x}{L/2-x}\right]\right) = f'(x) + \frac{4}{\pi E^*}\frac{P_s}{L}\operatorname{artanh}\left(\frac{2x}{L}\right). \tag{1.53}$$

Substitution of Eq. (1.53) in Eq. (1.49) gives the following expression for the contact pressure, which takes into account the elastic interaction of asperities:

$$p(x) = \frac{E^*}{2\pi}\sqrt{a^2 - x^2}\int_{-a}^{a}\left[f'(\xi) + \frac{4}{\pi E^*}\frac{P_s}{L}\operatorname{artanh}\left(\frac{2\xi}{L}\right)\right]\frac{1}{\sqrt{a^2-\xi^2}}\frac{1}{\xi - x}d\xi. \tag{1.54}$$

The total load applied to each asperity, taking into account symmetry of the function $f(x)$, is determined directly using Eq. (1.54) (Staierman, 1949; Barber, 2018):

$$P = \frac{E^*}{2}\int_{-a}^{a}\frac{f'(\xi)\xi d\xi}{\sqrt{a^2-\xi^2}} + \frac{2}{\pi}\frac{P_s}{L}\int_{-a}^{a}\frac{\xi}{\sqrt{a^2-\xi^2}}\operatorname{artanh}\left(\frac{2\xi}{L}\right)d\xi. \tag{1.55}$$

Calculation of the integral in the second term of Eq. (1.55) with condition of $2a < L$ gives the following expression for the load:

$$P = \frac{E^*}{2}\int_{-a}^{a}\frac{f'(\xi)\xi d\xi}{\sqrt{a^2-\xi^2}} + \frac{P_s}{L}\left(L-\sqrt{L^2-4a^2}\right). \tag{1.56}$$

Using Eq. (1.56), it is possible to simplify the integration of the second term in square brackets in Eq. (1.54). For this purpose, the method based on the Abel transform of the function $\partial P/\partial a$ (Barber, 2018) was used. Then Eq. (1.54) is reduced to the following form (see also Goryacheva and Tsukanov, 2020b)

$$p(x) = \frac{1}{\pi}\int_{x}^{a}\frac{P'_a(\xi)d\xi}{\sqrt{\xi^2-x^2}} = \frac{E^*}{2\pi}\sqrt{a^2-x^2}\int_{-a}^{a}\frac{f'(\xi)d\xi}{\sqrt{a^2-\xi^2}(\xi-x)} + \frac{2P_s}{\pi L}\arctan\left(\frac{2\sqrt{a^2-x^2}}{\sqrt{L^2-4a^2}}\right), \tag{1.57}$$

where $P'_a = \partial P/\partial a$.

It should be noted that in the 2-D periodic problem, the effect of elastic interaction on the contact pressure is characterized by a function similar to that for 3-D axisymmetric asperities arranged at the sites of hexagonal lattice (see Eq. 1.45).

As an example, we consider a profile described by the function $f(x) = \Delta(1-\cos(2\pi x/L))$, where Δ, L are the amplitude and the period, respectively. The exact expressions for determining the contact pressure distribution and the dependence of the mean pressure $\bar{p}$ on contact zone half-width are presented by Eqs. (1.9) and (1.11). Fig. 1.10 illustrates the

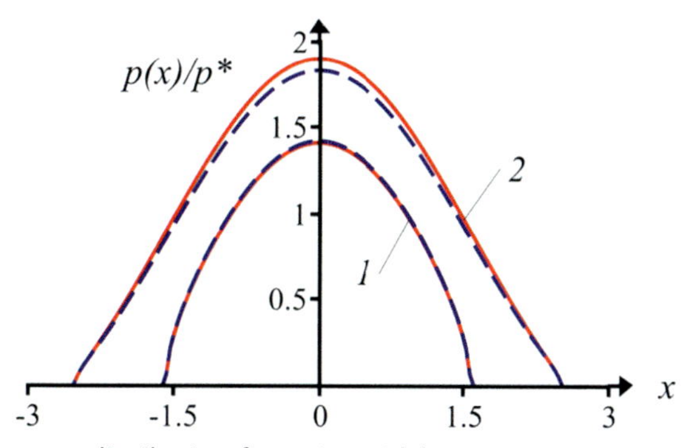

Figure 1.10 Pressure distribution for a sinusoidal waviness contacting with an elastic half-plane at $2a/L = 0.5$ (*1*) and $2a/L = 0.8$ (*2*): exact solution (*solid lines*), approximate solution based on the localization principle (*dashed lines*).

pressure distributions within a single contact spot for a sinusoidal waviness in contact with an elastic half-plane calculated from Eq. (1.9) and from Eq. (1.57).

The results indicate that the solution based on the localization principle allows predicting the distribution of contact pressures with sufficient accuracy up to high loads (high contact density); a significant discrepancy between the exact and approximate values of contact characteristics begins only at $2a/L \approx 0.7$. Note, that for such high values of contact density the solution for almost complete contact can be applied (Johnson, 1985).

We also compared the pressure distributions calculated for a periodic system of cylindrical (parabolic) asperities in which the radius R of curvature significantly exceeds their height. This problem was previously considered by Kuznetsov (1978). The relation between the half-width a of a contact zone and the total load per one asperity P_s has the form (Kuznetsov, 1978):

$$a = \frac{L}{\pi} \arccos\left(\exp\left(-\frac{2\pi P_s R}{L^2 E^*} \right) \right), \tag{1.58}$$

where L is the distance between the centers of the contact segments.

The approximate expressions for the pressure distribution and the total load on a single asperity, taking into account elastic interaction, follow from Eqs. (1.56) and (1.57) and Hertz solution for the cylinder penetrating into the elastic half-plane:

$$p(x) = \frac{E^*}{2R}\sqrt{a^2 - x^2} + \frac{E^* a^2}{2RL} \arctan\left(\frac{2\sqrt{a^2 - x^2}}{\sqrt{L^2 - 4a^2}} \right); \tag{1.59}$$

$$P_s = \frac{\pi E^* a^2}{4R}\left(2 - \frac{\sqrt{L^2 - 4a^2}}{L} \right). \tag{1.60}$$

The pressure distributions calculated by Goryacheva and Tsukanov (2020) for periodic system of cylindrical (parabolic) asperities with use of Eq. (1.59) in comparison with Hertz theory (in 2-D formulation) for two values of the nominal pressure are shown in Fig. 1.11.

The results indicate that due to asperity interaction the contact zone size decreases and the maximum value of the contact pressure increases. With increasing the nominal pressure the contact density a/L also increases, and the pressure profile becomes significantly different from the Hertzian one. This result is in good agreement with the 3-D analysis for spherical asperities performed in Section 1.3.3.

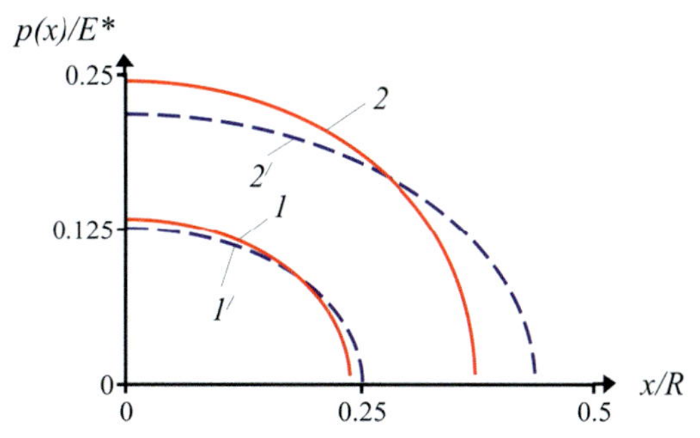

Figure 1.11 The contact pressure distribution under the cylindrical asperity for the periodic system (1, 2) and from the Hertz solution (1′, 2′) at $\overline{p}/E^* = 0.05$ (1, 1′) and $\overline{p}/E^* = 0.15$ (2, 2′).

The dependencies of the dimensionless contact half-width a/L on the dimensionless load for a periodic system of parabolic asperities calculated based on the exact and approximate solutions are presented in Fig. 1.12. The results indicate that the localization principle solution is close to the exact solution up to the value $a/L \approx 0.35$. The discrepancy with the Hertzian curve corresponding to noninteracting asperities begins at $a/L \approx 0.125$. For $a/L < 0.125$ the error of calculation using the Hertz theory is less than 3%. This result is similar to the case of a 3-D system of spherical asperities in contact with the elastic half-space (see Fig. 1.5). At the large

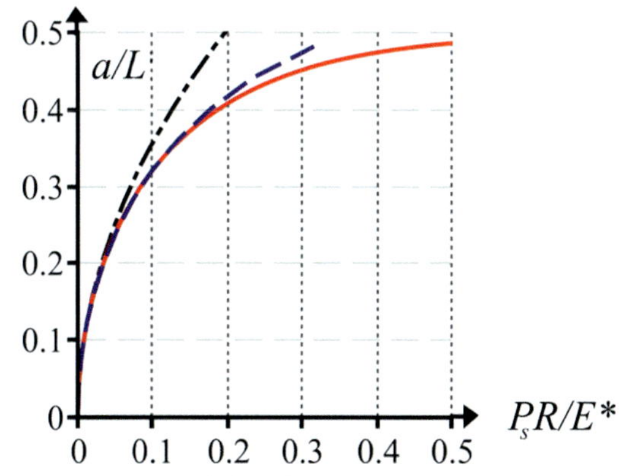

Figure 1.12 The dependence of the dimensionless contact half-width a/L on the dimensionless load for a periodic system of cylindrical asperities: exact solution (*solid line*), approximate solution based on the localization principle (*dashed line*), Hertz theory (*dash-dot line*).

values of load, the approximate solution overestimates the contact zone size in comparison with the exact solution (1.58).

Other examples of the localization method application in 2-D discrete contact problems (surface regular microgeometry is described by wedge profiles or parabolic asperities with double contact segments within a period) were given by Goryacheva and Tsukanov (2020b). The results obtained show that the application of the localization method in a plane contact problem for bodies with a periodic regular microrelief allows to calculate the contact characteristics with high accuracy up to large contact densities ($2a/L \approx 0.7$). For all considered cases, the increase in asperities density leads to an increase in peak pressure, and also to the reduction of contact half-width at fixed nominal pressure. Qualitatively, the form of contact pressure distribution depends on the shape of asperities.

1.4 Contact problems with bounded nominal contact region

The periodic contact problems considered above are characterized by uniform distribution of the nominal contact pressure over the boundary of the elastic half-space. Within each period, the load distribution between contact spots depends only on the heights of indenters and variations in contact density.

For a finite number of indenters interacting with an elastic half-space, the nominal contact region is bounded. In this case, the load distribution between the indenters is nonuniform even though all indenters have the same height and they are arranged uniformly within a bounded nominal contact region.

In this section, the solutions of the contact problem for a finite number of indenters (with various shapes of contacting surface) penetrating into the elastic half-space are presented. These solutions make it possible to analyze the dependence of the contact characteristics (load distribution, real contact area, etc.) on the spatial arrangement of the indenters.

1.4.1 Contact problem formulation

The contact interaction of a system of N identical indenters with an elastic half-space is considered (Fig. 1.13).

The shape of the contacting surface of each indenter is described by the function $f_i(r)$ in the system of coordinates related to the axis of the indenter, which is perpendicular to the undeformed surface of the half-space. The

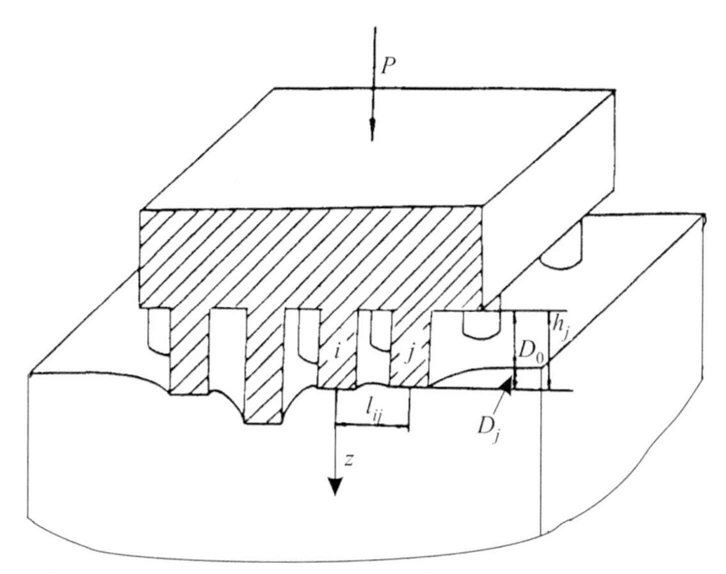

Figure 1.13 Scheme of contact of the system of indenters and an elastic half-space.

distance l_{ij} between the axis of symmetry of the i-th and j-th indenters as well as the heights h_i of the indenters are known.

In loading, the real contact area of this system with the elastic half-space boundary consists of a set of subregions ω_i ($i = 1, 2, ..., N$). The remaining boundary of the half-space is stress free.

We introduce the coordinate system $Oxyz$ whose Oz-axis coincides with the axis of revolution of an arbitrary fixed i-th indenter, and the plane Oxy coincides with the undeformed half-space surface.

Let us formulate thc boundary conditions for the i-th indenter and replace the action of the other indenters on the boundary of the elastic half-space by the corresponding pressure distributed over the aggregate region $\bigcup_{\substack{j=1 \\ j\neq i}}^{N} \omega_j$. The elastic displacement $u_z^{i1}(x, y)$ of the half-space surface in the z-axis direction within the subregion ω_i caused by the pressure $p_j(x, y)$, $(x, y) \in \omega_j$, $(j = 1, 2, ..., N,\ i \neq j)$ is calculated from Boussinesq's solution:

$$u_z^{i1}(x, y) = \frac{1 - \nu^2}{\pi E} \sum_{\substack{j=1 \\ j\neq i}}^{N} \iint_{\omega_i} \frac{p_j(x', y')dx'dy'}{\sqrt{\left(X_j + x' - x\right)^2 + \left(Y_j + y' - y\right)^2}}.$$

Here (X_j, Y_j) are the coordinates of the central point of the subregion ω_j.

The pressure $p_j(x, y)$ is not known in advance. To simplify the problem, we approximate $u_z^{i1}(x, y)$ by the following function:

$$u_z^{i1}(x, y) = \frac{1 - \nu^2}{\pi E} \sum_{\substack{j=1 \\ j \neq i}}^{N} \iint_{\omega_j} \frac{P_j dx' dy'}{\sqrt{(X_j - x)^2 + (Y_j - y)^2}} \tag{1.61}$$

where P_j is the concentrated force, $P_j = \iint_{\omega_j} p_j(x, y) dx dy$, which is applied at the center (X_j, Y_j) of the subregion ω_j. The accuracy of this approximation can be estimated for the particular case of the axially symmetric function $p_j(x', y') = p(r)$ $(r \leq a)$:

$$\int_0^{2\pi} \int_0^a \frac{p(r) r dr d\theta}{\sqrt{r^2 + l^2 - 2rl \cos\theta}} - \frac{P}{l} = \int_0^a p(r) \left[\frac{4}{r + l} K\left(\frac{2\sqrt{rl}}{r + l} \right) - \frac{2\pi}{l} \right] r dr$$

$$= 4 \int_0^a p(r) \frac{r}{l} \left[K\left(\frac{r}{l} \right) - \frac{\pi}{2} \right] dr = \frac{\pi}{2} \int_0^a p(r) \left[\left(\frac{r}{l} \right)^3 + O\left(\left(\frac{r}{l} \right)^5 \right) \right] dr$$

$$\leq \frac{P}{4} \left(\frac{a^2}{l^3} + O\left(\frac{a^4}{l^5} \right) \right), \quad l = \sqrt{(X_j - x)^2 + (Y_j - y)^2}, \quad P = 2\pi \int_0^a p(r) r dr.$$

Here $K(x)$ is the elliptic integral of the first kind. The following relations have been used to obtain this estimation:

$$\int_0^{\pi} \frac{d\theta}{\sqrt{r^2 + l^2 - 2rl \cos\theta}} = \frac{2}{r + l} K\left(\frac{2\sqrt{rl}}{r + l} \right), \quad K\left(\frac{2\sqrt{x}}{1 + x} \right) = (1 + x) K(x) \tag{1.62}$$

The superposition principle, which is valid for the linear theory of elasticity, makes it possible to present the displacements of the boundary of the elastic half-space along the axis Oz under the i-th punch, as a sum of the displacement $u_z^{i1}(x, y)$ (1.61) and the elastic displacement $u_z^{i2}(x, y)$ due to the pressure $p_i(x, y)$ distributed over the i-th punch base within the subregion ω_i. As a result, the pressure $p_i(x, y)$ can be determined from the solution of the following problem for the elastic half-space with the mixed boundary conditions:

$$
\begin{aligned}
&u_z^{i1}(x,y) + u_z^{i2}(x,y) = D_i - f_i\left(\sqrt{x^2+y^2}\right),\\
&\tau_{xz} = \tau_{yz} = 0, \quad (x,y)\in\omega_i,\\
&\sigma_z = \tau_{xz} = \tau_{yz} = 0, \quad (x,y)\notin\omega_i,
\end{aligned}
\tag{1.63}
$$

where D_i is the displacement of the i-th punch along the z-axis.

In some applications, it is essential to determine the load distribution between the indenters in the system penetrating into the elastic half-space at given depth D_0. Using Betti's theorem, the following relation has been derived for the i-th punch of the system (Goryacheva, 1998):

$$
P_i = \frac{E}{\pi(1-\nu^2)} \int_0^{2\pi} \int_0^{a_i} \frac{u_z^{i2}(r,\theta)\, r\, dr\, d\theta}{\sqrt{a_i^2 - r^2}}. \tag{1.64}
$$

Substituting the function $u_z^{i2}(x, y)$ from Eq. (1.63) in Eq. (1.64) and taking into account expression (1.61) for the function $u_z^{i1}(x, y)$, we obtain the following relation between the load P_i and displacement D_i of the i-th punch of the system taking into account the additional loading P_j $(j = 1, 2, \ldots, N, j \neq i)$ of the other punches of the system (Goryacheva, 1998):

$$
P_i = \frac{2E}{1-\nu^2} \int_0^{a_i} (D_i - f(r)) \frac{r\,dr}{\sqrt{a_i^2 - r^2}} - \frac{2}{\pi} \sum_{\substack{j=1\\ j\neq i}}^{N} P_j \arcsin\frac{a_i}{l_{ij}}, \tag{1.65}
$$

where $l_{ij} = \sqrt{X_j^2 + Y_j^2}$. In derivation of Eq. (1.65), we used Eq. (1.62) and also the following relations:

$$
\int_0^{x^2} \frac{dx'}{\sqrt{(x^2 - x')(1 - x'\sin^2\phi)}} = \frac{1}{\sin\phi} \ln\frac{1 + x\sin\phi}{1 - x\sin\phi}, \quad (|x| < 1),
$$

$$
\int_0^{\pi/2} \frac{1}{\sin\phi} \ln\frac{1 + x\sin\phi}{1 - x\sin\phi}\, d\phi = \pi \arcsin x.
$$

Taking into account relations (1.65) for each punch of the system and the contact condition

$$D_i = h_i - D_0, \tag{1.66}$$

we get 2N equations for calculating the values of loads P_i and penetrations D_i $(i = 1, 2, ..., N)$.

If the approach of the bodies D_0 is not given in advance, and we know the load P applied to the system of punches (see Fig. 1.13), then we add to Eqs. (1.65) and (1.66) the following equilibrium condition:

$$\sum_{i=1}^{N} P_i = P. \tag{1.67}$$

If the contacting surfaces of the axially symmetric punches are smooth, the radius of each contact spot is also an unknown value. It can be found from the condition

$$p(a_i, \theta) = 0, \quad 0 < \theta \leq 2\pi.$$

From this relation and also from the equilibrium condition

$$P_i(a_i) = \int_0^{a_i} \int_0^{2\pi} p(r, \theta) r dr d\theta,$$

it follows that $\frac{\partial P_i}{\partial a_i} = 0$. Differentiation of Eq. (1.65) with respect to a_i gives

$$D_i = -a_i \int_0^{a_i} \frac{f'(r) dr}{\sqrt{a_i^2 - r^2}} + \frac{1 - \nu^2}{\pi E} \sum_{\substack{j=1 \\ j \neq i}}^{N} \frac{P_j}{\sqrt{l_{ij}^2 - a_i^2}}. \tag{1.68}$$

Eqs. (1.65)–(1.68) provide the complete system of equations to determine the values of D_i, a_i, and P_i for a system of punches, the shapes of which are described by a continuously differentiable function.

1.4.2 Contact problem for a system of spherical punches

For punches with a spherical contacting surface of radius R, i.e., $f(r) = r^2/(2R)$, and given spatial arrangement (given values h_i and l_{ij}), Eqs. (1.65)–(1.68), under the assumption that the contact region of each punch and the elastic half-space is closed to a circle area, take the form

$$
\begin{aligned}
&h_i - D_0 = \frac{a_i^2}{R} + \frac{1-\nu^2}{\pi E}\sum_{\substack{j=1\\ j\neq i}}^{N}\frac{P_j}{\sqrt{l_{ij}^2 - a_i^2}}, \quad P_i = \frac{4Ea_i^3}{3R(1-\nu^2)}\\
&+\frac{2}{\pi}\sum_{\substack{j=1\\ j\neq i}}^{N} P_j\left[\frac{a_i}{\sqrt{l_{ij}^2 - a_i^2}} - \arcsin\frac{a_i}{l_{ij}}\right], \quad P = \sum_{i=1}^{N} P_i, \quad i = 1, 2, ..., N.
\end{aligned}
\tag{1.69}
$$

System (1.69) specifies the distribution of forces P_i among N punches, which are loaded with the total force P and interact with the elastic half-space, the radii a_i of the contact subregions ω_i, the total real area of contact $A_r = \pi\sum_{i=1}^{N} a_i^2$, and the approach-load dependence $D_0(P)$. It follows from the second group of Eq. (1.69) that the radius of the i-th contact spot can be determined with accuracy of order $\left(a_i/l_{ij}\right)^3$ by the Hertz formula:

$$
a_i = \sqrt[3]{\frac{3(1-\nu^2)RP_i}{4E}}.
$$

Then the real area of contact can be approximated by the formula

$$
A_r = \pi\sum_{i=1}^{N}\left[\frac{3(1-\nu^2)RP_i}{4E}\right]^{2/3},
$$

where P_i is determined from the first group of Eq. (1.69).

Let us consider the system of $N = 52$ spherical punches of radius R, located at the same height and distributed at the sites of a square lattice (l is the lattice pitch). Curve 1 in Fig. 1.14 shows the dependence of the relative area of contact A_r/A_a (A_a is the nominal area of contact) versus the pressure $\overline{p} = P/A_a$ calculated from Eq. (1.69) for $l/R = 0.5$.

Curve 2 in Fig. 1.14 is calculated using the Hertz theory and neglecting the redistribution of the loads applied to each contact spot due to the interaction between contact spots. As follows from the results of calculations for the chosen relative lattice pitch $l/R = 0.5$, starting from $A_r/A_a = 0.3$, there is a noticeable error in the calculation of the real area of contact from the theory, which ignores interaction between contact spots.

Fig. 1.15 illustrates the dependence of the depth of penetration D upon the load P for the system of spherical indenters characterized by various distance between indenters (parameter l/R).

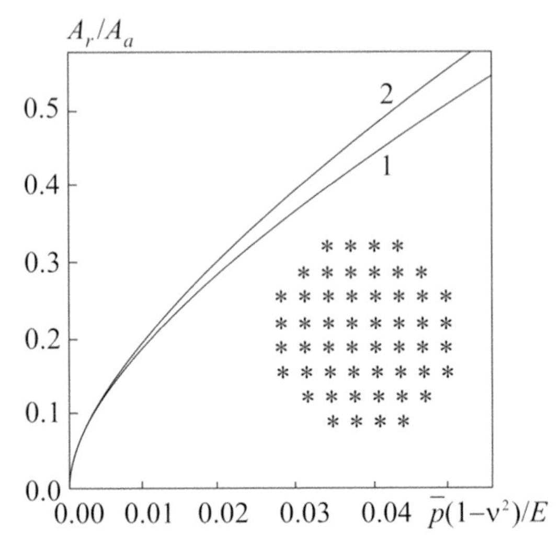

Figure 1.14 The dependence of the relative area of contact upon nominal pressure at $l/R = 0.5$ calculated from the multiple contact model (curve 1) and the Hertz model (curve 2).

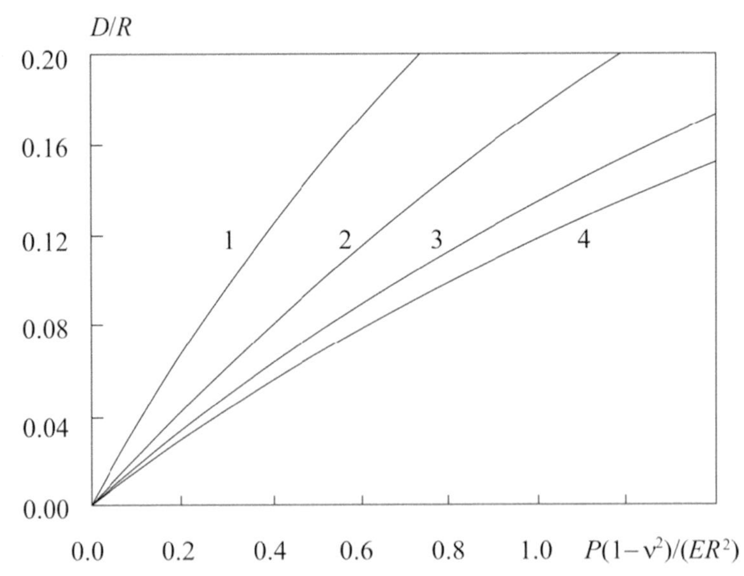

Figure 1.15 Dependencies of the dimensionless depth of penetration upon the dimensionless load for the various values of the parameter l/R: $l/R = 0.5$ (curve 1), $l/R = 1$ (curve 2), $l/R = 1.5$ (curve 3), and $l/R = 2$ (curve 4).

The results indicate that the higher is the contact density (i.e., the smaller is the parameter l/R), the smaller is the load required to achieve the given depth of penetration. Similar results were obtained theoretically and experimentally when studying the interaction of a system of cylindrical punches, located at the same level, with an elastic half-space (Goryacheva and Dobychin, 1991; Goryacheva, 1998).

From the results of the analysis, we conclude that the calculation methods that do not take into account the interaction of the contact spots give overestimated values for the contact stiffness dP/dD and the real area of contact A_r; the error increases with increasing the number of contacts and their density. The geometrical imperfections of a surface, in particular its waviness and distortion, lead to the localization of contact spots within the so-called contour regions where the high densities of contact spots occur. So even a moderate load provides a high relative contact area within the contour regions, and the error of calculation based on the simplified theory can be large.

1.5 Conclusion

In this chapter the unified approximate analytical method to solve 2-D and 3-D periodic contact problems for body with a regular surface micro-geometry penetrating the elastic half-space (half-plane) is developed. This method based on the localization principle makes it possible to take into account the interaction between contact spots. The equations derived for 3-D and 2-D cases show identical structure for different dimensionality of the problem. The advantage of the approach based on the method of localization is the ability to separately consider the effects associated with the shape of asperities and the relative spacing between them. The method allows us to simplify the calculation of contact characteristics for complex-shaped regular texture, for which a straightforward analytical solution does not exist. The approach developed can be used for the solution of the 2-D and 3-D periodic contact problems with complicated boundary conditions and also for the other models of the deformable bodies. The application of this method to study the contact characteristics and their evolution in time in contact of the periodic system of spherical indenters and a viscoelastic half-space is developed in Yakovenko and Goryacheva (2021). The example of calculation of contact characteristics in normal discrete contact with adhesion of different nature at the interface is presented in Chapter 2.

For the contact interaction with bounded nominal contact region, the contact problem for a finite system of punches penetrating into an elastic half-space was also studied. The results make it possible to analyze the effects of contact spots interaction and the bounded nominal contact region on the real contact area, load distribution within the nominal contact region, and the approach of the contacting bodies under given total load applied to the system of indenters.

References

Antipov, Y.A., Arutyunyan, N.K., 1991. Contact problems of the theory of elasticity with friction and adhesion. J. Appl. Math. Mech. 55 (6), 887–901.

Barber, J.R., 2018. Contact Mechanics. Springer International Publishing.

Bartenev, G.M., Lavrentiev, V.V., 1972. Friction and Wear of Polymers, (I.R.). Khimiya, Moscow.

Block, J.M., Keer, L.M., 2008. Periodic contact problems in plane elasticity. J. Mech. Mater. Struct. 3, 1207–1237.

Boussinesq, J., 1885. Application des Potentials a l'Etude de l'Equilibre et du Mouvement des Solides Elastiques. Gauthier-Villars, Paris.

Chekina, O.G., Keer, L.M., 1999. A new approach to calculation of contact characteristics. Trans. of ASME, J.Tribol., V. 121, 20–27.

Demkin, N.B., 1970. Contact of Rough Surfaces, (I.R.). Nauka, Moscow.

Dundurs, J., Tsai, K.C., Keer, L.M., 1973. Contact between elastic bodies with wavy surfaces. J. Elasticity 3, 109–115.

Galin, L.A., 1953. Contact Problems of Theory of Elasticity, (I.R.). Gostekhizdat, Moscow (English translation by Moss, H., North Carolina State College, Department of Mathematics, 1961), (Chapter 2).

Galin, L.A., Gladwell, G.M.L. (Eds.), 2008. Contact Problems. Springer, p. 315.

Goryacheva, I.G., 1998. Contact Mechanics in Tribology. Kluwer Acad.Publ., p. 344

Goryacheva, I.G., Dobychin, M.N., 1991. Multiple contact model in the problems of tribomechanics. Tribol. Int. 24 (1), 29–35.

Goryacheva, I.G., Torskaya, E.V., 2003. Stress and fracture analysis in periodic contact problem for coated bodies. Fatig. Fract. Eng. Mater. Struct. 26 (4), 343–348.

Goryacheva, I.G., Torskaya, E.V., 2019. Contact of multi-level periodic system of indenters with coated elastic half-space. Facta Univ. – Ser. Mech. Eng. 17 (2), 149–159.

Goryacheva, I.G., Tsukanov, I.Y., 2020a. Development of discrete contact mechanics with applications to study the frictional interaction of deformable bodies. Mech. Solid. 55 (8), 1441–1462.

Goryacheva, I.G., Tsukanov, I.Y., 2020b. Analysis of elastic normal contact of surfaces with regular microgeometry based on the localization principle. Front. Mech. Eng. 6, 1–10.

Greenwood, J.A., Williamson, J.B.P., 1966. Contact of nominally flat surfaces. Proc.Roy.Soc., A 295 (1442), 300–319.

Johnson, K.L., 1985. Contact Mechanics. Cambridge University Press.

Johnson, K.L., Greenwood, J.A., Higginson, J.G., 1985. The contact of elastic wavy surfaces. Intern. J.Mech. Sci. 27 (6), 383–396.

Kragelsky, I.V., Dobychin, M.N., Kombalov, 1982. Friction and Wear: Calculation Methods. Pergamon Press, Oxford.

Kuznetsov, Y.A., 1976. Periodic contact problem for half-plane allowing for forces of friction. Sov. Appl. Mech. 12, 1014–1019.

Kuznetsov, Y.A., 1978. The use of automorphic functions in the plane theory of elasticity. Mech. Solid. 6, 35–44.
Kuznetsov, Y.A., Gorohovskii, G.A., 1977. On real contact pressure. Probl.Treniya Iznashivaniya 12, 10–13.
Kuznetsov, Y.A., Gorohovskii, G.A., 1978. Effect of roughness on the stress state of bodies in frictional contact. Appl. Mech. 14 (9), 62–68.
Kuznetsov, Y.A., Gorohovskii, G.A., 1981. Effect of tangential load on the stressed state of rubbing rough bodies. Wear 73 (1), 41–58.
Lubrecht, A.A., Ioannides, E.V., 1991. A fast solution of dry contact problem and the associated subsurface stress field, using multilevel techniques. ASME. J.Tribol. 113, 128–133.
Muskhelishvili, N.I., 1953. Some Basic Problems of the Mathematical Theory of Elasticity. Noordhoff, Holland.
Rostami, A., Jackson, R.L., 2013. Predictions of the average surface separation and stiffness between contacting elastic and elastic-plastic sinusoidal surfaces. Proc. Inst. Mech. Eng., Part J. 227, 1376–1385.
Sadowsky, M., 1928. Zweidimensionale probleme der Elastizitatstheorie. Z. Angew. Math. Mech. 8 (2), 107–121.
Staierman, I.Y., 1949. Contact Problem of the Elasticity Theory. (I.R.). Gostekhizdat, Moscow.
Stanley, H.M., Kato, T., 1997. An FFT-based method for rough surface contact. Trans. ASME. J.Tribol. 119, 481–485.
Timoshenko, S., Goodier, J.N., 1951. Theory of Elasticity, third ed. McGraw-Hill, New York-London.
Tsukanov, I.Y., 2018a. Periodical contact problem for a surface with two-level waviness. Mech. Solid. 53 (1), 129–136.
Tsukanov, I.Y., 2018b. Partial contact of a rigid multisinusoidal wavy surface with an elastic half-plane. Adv. Tribol. 8431467.
Tsukanov, I.Y., 2019. An extended asymptotic analysis for elastic contact of three-dimensional wavy surfaces. Tribol. Lett. 67 (4), 107–113.
Westergaard, H.M., 1939. Bearing pressures and cracks. J. Appl. Mech.-T. ASME. 6, 49–52.
Whitehouse, D.J., 1994. Handbook of Surface Metrology. Inst.of Physics, Bristol.
Yakovenko, A.A., Goryacheva, I.G., 2021. The periodic contact problem for spherical indenters and viscoelastic half-space. Tribol. Int. 161, 107078.
Yastrebov, V.A., Anciaux, G., Molinari, J.-F., 2014. The contact of elastic regular wavy surfaces revisited. Tribol. Lett. 56, 171–183.

CHAPTER 2

Effect of adhesion in normal discrete contact

In this chapter, models are presented to calculate the distributions of real contact pressure and displacement in the contact interface between elastic bodies with nominally flat surfaces, one of which has a regular microrelief, taking into account adhesion of different nature — molecular attraction or capillary adhesion due to the formation of fluid menisci near real contact spots.

2.1 Adhesion of a different nature: molecular and capillary

The periodic contact problem formulations considered in Chapter 1 assume the action of compressive (positive) stresses in the contact region and zero stresses outside it, the contact stress being equal to zero at the boundary of the contact region. It is known however that real solids and fluid films covering their surfaces possess surface energy which lead to adhesive attraction and, as a consequence of that, tensile (negative) stresses acting both in the contact region and outside it.

In this chapter, two types of adhesion are considered — molecular and capillary. They both are caused by the action of intermolecular forces in the interacting solids and films of fluid covering them. To distinguish between two important cases of action of the intermolecular forces in contact, we will use the term **molecular adhesion** for adhesion of dry surfaces caused by the action of van der Waals intermolecular forces between the molecules of two solids. The term **capillary adhesion** will be used to describe the attraction caused by the capillary forces in menisci of fluid in the gap between surfaces. We will assume that in the latter case, the intermolecular van der Waals forces between solid surfaces acting through liquid medium are insignificant in comparison with the capillary forces.

Discrete Contact Mechanics with Applications in Tribology
ISBN 978-0-12-821799-3
https://doi.org/10.1016/B978-0-12-821799-3.00004-2

2.1.1 Molecular adhesion

Phenomenological relation of the energy of intermolecular interaction and distance between molecules was suggested by Lennard-Jones. Under the assumption that the force of interaction of two molecules depends on the distance between them and is directed along the line connecting them, the potential of interaction has the form (Israelachvili, 1992):

$$U_0(r) = -\frac{A_1}{r^6} + \frac{A_2}{r^{12}}, \tag{2.1}$$

where r is the distance between the centers of the interacting molecules, and A_1 and A_2 are constants. The first term in Eq. (2.1) corresponds to the attraction of molecules, while the second one to the repulsion.

The potential of adhesive interaction $U_a(h)$ between two rigid half-spaces with parallel boundaries as a function of distance between them h is a result of the interaction of all molecules of one half-space with all molecules of the other half-space that can be determined from Eq. (2.1) by the double integration (Maugis, 2000):

$$U_a(h) = -\frac{4w_a}{3}\left(\left(\frac{z_0}{h}\right)^2 - \frac{1}{4}\left(\frac{z_0}{h}\right)^8\right). \tag{2.2}$$

From this, the adhesion pressure $p_a(h) = \partial U_a(h)/\partial h$ acting on the surface is determined by

$$p_a(h) = -\frac{8w_a}{3z_0}\left(\left(\frac{z_0}{h}\right)^3 - \left(\frac{z_0}{h}\right)^9\right), \tag{2.3}$$

where z_0 is the equilibrium distance at which $p_a(z_0) = 0$, and w_a is the specific work of adhesion defined as the work per unit area done to separate the surfaces from the equilibrium distance to infinity:

$$w_a = \int_{z_0}^{\infty} p_a(h)dh \tag{2.4}$$

The adhesion pressure as a function of the distance (gap) between surfaces Eq. (2.3) is shown in Fig. 2.1 (curve 1). If the gap is smaller than the equilibrium distance, $h < z_0$, the pressure is positive that corresponds to the repulsion of surfaces. For $h > z_0$, the adhesive attraction occurs, the adhesive pressure is negative, and its absolute value attains maximum for a certain value of the gap and tends to zero as h further increases.

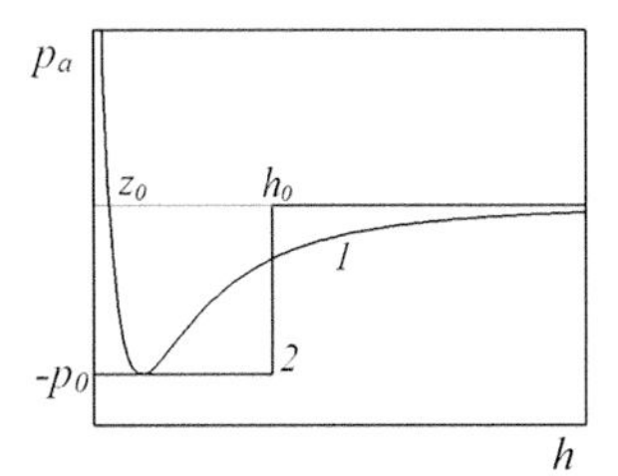

Figure 2.1 Adhesion pressure as a function of the gap between two plane parallel surfaces of rigid bodies.

In the case of deformable interacting bodies with curved surfaces, the solution of adhesive contact problem is a challenging task due to mutual influence of the surface displacements and adhesive forces. To overcome this difficulty, approximate models were developed for the description of the adhesive contact of elastic bodies, in particular, the adhesive contact between a sphere and a half-space or between two spheres. The JKR model (Johnson et al., 1971) is based on calculating the energy balance and considers changes in both elastic energy due to the contact deformation of interacting bodies and surface energy as a result of change in the contact area. According to the DMT model (Derjaguin et al., 1975), the intermolecular forces act in a certain region around the contact region and their value depends on the value of gap, the contact stress distribution coinciding with the Hertzian one. Both the JRK and DMT simplified models adequately describe the adhesion contact in two limiting cases, which can be distinguished from each other by the Tabor parameter. This parameter relates the effective elastic modulus E^* of the interacting spheres to their surface properties and effective radius R (Tabor, 1977):

$$\mu_T = \left(\frac{R w_a^2}{E^{*2} z_0^3} \right)^{1/3} \tag{2.5}$$

For $\mu_T << 1$ (relatively hard spheres of small radii and low surface energy) the DMT model is applicable, whereas for $\mu_T >> 1$ (soft materials, large radii, and high surface energy) the JKR model is valid with high accuracy (Tabor, 1977).

The JKR model was generalized for the body shapes described by the power law function (Borodich et al., 2014a) and for anisotropic bodies (Barber and Ciavarella, 2014; Borodich et al., 2014b).

The advantage of the JKR and DMT models is their simplicity; however, they cannot be used in the entire range of the elastic and adhesive properties of solids. To obtain the exact solution of an adhesion contact problem, it must be considered in the accurate formulation taking into account the influence of the adhesion forces on the stress-strain state of interacting bodies. In doing so, one should consider the adhesive interaction of all molecules of one body to all molecules of the other body in accordance with the potential of interaction Eq. (2.1). To simplify the formulation, Derjaguin has suggested an approximation, according to which the adhesion forces are applied only to the boundaries of interacting bodies (Derjaguin, 1934). At each point, these boundaries are considered a pair of parallel surfaces, and the pressure applied to them due to adhesion is taken equal to the adhesion pressure between two plane parallel boundaries of rigid half-spaces (Fig. 2.2). Within the framework of the Derjaguin approximation, the models using various functions of adhesion pressure versus distance were developed, based on the Lennard-Jones pressure Eq. (2.3) as well as on its approximations by various functions.

Among such models, the Maugis-Dugdale model (Maugis, 1992) is particularly interesting, since it combines relative simplicity with taking into consideration mutual influence of adhesion forces and contact deformations (self-consistent approach). According to this model, the adhesion pressure as a function of the gap between surfaces is specified as follows (Fig. 2.1, curve 2):

$$p_a(h) = \begin{cases} -p_0, & 0 < h \le h_0 \\ 0, & h > h_0 \end{cases} \tag{2.6}$$

In this model, the contact stress distribution and the contact radius differ from the corresponding Hertzian values, since they are influenced by the adhesion pressure acting outside the contact region. In this case, the specific work of adhesion is defined by

$$w_a = \int_0^{\infty} p_a(h)dh = p_0 h_0. \tag{2.7}$$

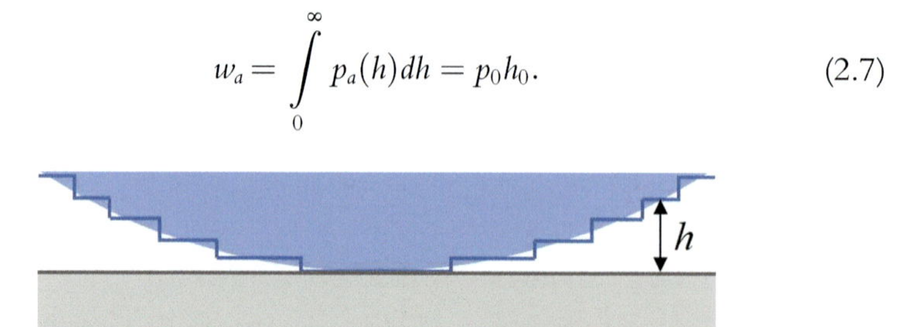

Figure 2.2 Scheme of the Derjaguin approximation.

For a spherical punch in contact with an elastic half-space, an analytic solution was obtained, which, unlike the JKR and DMT models, is applicable for bodies of any rigidity and parameters of adhesion (Maugis, 1992). Johnson and Greenwood (1997) suggested an adhesion map to choose a suitable model of interaction between elastic spheres under specified parameters of loading and adhesion: the JKR, DMT, Maugis-Dugdale model, or the Hertz solution for the cases where the effect of adhesion on the contact characteristics is insignificant.

The Maugis-Dugdale model was generalized for the case of a punch whose shape is described by the power law function of an even degree (Goryacheva and Makhovskaya, 2001). Based on this result, the solution was constructed for a piecewise approximation of the adhesive potential of an arbitrary form (Goryacheva and Makhovskaya, 2004; Makhovskaya, 2016). This solution was obtained for the cases of both direct contact between surfaces and separated surfaces interacting via adhesive forces, which made it possible to analyze the contact characteristics in the entire range of loads. The Maugis-Dugdale model was extended to the case where an elastic half-space is in contact with an axisymmetric punch whose shape is described by the power law function of an arbitrary degree (Wei and Zhao, 2004; Zhijun and Jilin, 2007).

The models mentioned above include the concept of contact region inside which the contact condition—the equality of the displacements of two surfaces—is satisfied, while the adhesion pressure is applied outside the contact area and is self-consistently related to the elastic displacements in this region. However, the strict self-consistent approach assumes that there is no contact region in the traditional understanding of contact mechanics, but there is always a gap between surfaces. As a boundary condition, Eq. (2.3) or similar relations are used, which define both attraction and repulsion of surfaces, depending on the value of gap between them. The problem in such formulation was first considered and solved numerically by Greenwood (Greenwood, 1997). A similar approach was used (Soldatenkov, 2012) to develop the iterative method for the solution of adhesive contact problems. Solutions without using the Derjaguin approximation were also obtained, in which the interaction of molecules in subsurface layers was taken into account (Soldatenkov, 2013), including the case of coated bodies (Soldatenkov, 2016). These approaches require considerable numerical calculations.

In the present paper, we use the restricted self-consistent approach that assumes the existence of a contact area and adhesion forces acting outside this area (Hughes and White, 1979). This makes it possible to obtain analytic expressions for the contact pressures, elastic displacement of the boundaries of interacting bodies, and other contact characteristics.

2.1.2 Capillary adhesion

In fluids, intermolecular forces cause surface effects such as wetting and formation of drops and menisci. Films of fluid covering surfaces of solids can form menisci that serve as fluid bridges between interacting surfaces. This effect is called the capillary adhesion. In accordance with the Laplace formula, the pressure under a curved surface of a fluid meniscus is lower than the atmospheric pressure by the value

$$p_0 = \gamma_0\left(R_1^{-1} + R_2^{-1}\right), \tag{2.8}$$

where γ_0 is the surface tension of fluid, and R_1, R_2 are curvature radii of the meniscus in two mutually orthogonal directions. These radii are defined by the geometry of interacting surfaces, distance between them, and the boundary wetting angles θ_1, θ_2 of the two surfaces by fluid.

Let two elastic spheres be in contact over the region $r \leq a$, a meniscus being placed in the ring-shaped region $a \leq r \leq b$ between them (Fig. 2.3). There are two capillary forces acting between the surfaces. The first one is a force due to the surface tension of the fluid film at the outer boundary $r = b$ of the meniscus:

$$F_s = -2\pi b\gamma_0\left(\cos\theta_1 + \cos\theta_2\right). \tag{2.9}$$

The second one is the force caused by the Laplace pressure Eq. (2.8) acting in the region occupied by the meniscus:

$$F_L = -\pi\left(b^2 - a^2\right)p_0. \tag{2.10}$$

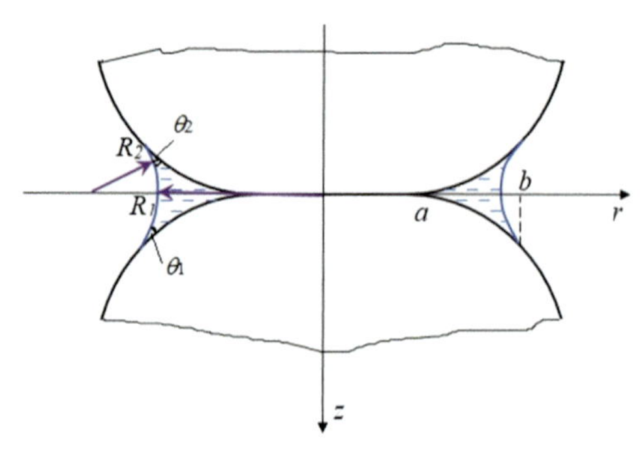

Figure 2.3 Scheme of capillary adhesion between two elastic spheres.

The total interaction force between the spheres is a sum of capillary adhesion forces Eqs. (2.9) and (2.10), and an elastic repulsion force acting in the contact region. The contact radius, as well as the geometry of the meniscus, are to be determined as a result of the contact problem solution taking into account the action of the capillary forces. Such formulation is similar to the self-consistent formulation of contact problems taking into account molecular adhesion of dry surfaces. A contact problem in such formulation cannot be solved analytically, so either numerical or approximate approaches are used.

Simplified models of the capillary adhesion between rigid bodies were suggested in the literature (Rabinowicz, 1965; Gao et al., 1995). The force of capillary adhesion between two rigid spheres as a function of distance between them was determined (Rabinovich et al., 2005; Megias-Alguacil and Gauckler, 2009; Payam and Fathipour, 2011), taking into account the exact geometry of the interface and under the condition that the volume of fluid in the meniscus remains constant in the process of interaction.

The solution obtained for the case of molecular adhesion was extended to the case of the interaction of a spherical punch with an elastic half-space in the presence of a fluid meniscus between them (Maugis and Gauthier-Manuel, 1994). The solution was constructed under the assumption of full wetting of the contacting surfaces by the fluid, i.e., for zero force of surface tension Eq. (2.9) and with no conditions imposed on the volume of the fluid. The solution of the contact problem with capillary adhesion for a constant volume of fluid in the meniscus was obtained (Goryacheva and Makhovskaya, 1999; Makhovskaya and Goryacheva, 1999) under the condition of full wetting and for a punch whose shape is described by a power law function. Fan and Gao (2001) considered the contact problem with a meniscus in 2-D formulation taking into account the force of surface tension of the fluid film. A series of numerical solutions was also constructed (see, for example, Butt et al., 2010; Zakerin et al., 2013).

2.2 Approach to study the normal discrete adhesive contact

Since the adhesion forces—both molecular and capillary—depend not only on the properties of interacting surfaces but also on the value of gap between them, the surface microgeometry has a considerable effect on the adhesive interaction. Applying a microrelief on a surface allows one to change the adhesive properties of this surface and either increase or decrease the adhesion (Briggs and Briscoe, 1977; Purtov et al., 2013).

The solution of a 2-D contact problem of adhesive interaction between a periodic wavy surface and an elastic half-space was obtained for an approximate formulation based on the JKR model (Johnson, 1995) and by using the Maugis-Dugdale model (Hui et al., 2001; Jin et al., 2016). Numerical solutions of the plane periodic contact problem were obtained for the Maugis-Dugdale (Adams, 2004) and Lennard-Jones (Wu, 2012) potentials.

A method of the 3-D periodic contact problem solution was suggested (Makhovskaya, 2003; Goryacheva and Makhovskaya, 2008) based on the method of localization for the cases of both molecular and capillary adhesion. In what follows, we present this method and some results of the discrete contact problem analysis.

2.2.1 Problem formulation

Let an elastic half-space interact with a periodic system of axisymmetric punches of the same height located at the nodes of the hexagonal lattice with a pitch l (Fig. 2.4a).

The origin of the local cylindrical system of coordinates coincides with the point at which the half-space touches a punch before deformation. The z-axis is directed inward to the elastic half-space. In this system of coordinates, the shape of each punch is described by the function $f(r) = Ar^{2n}$, where n is an integer.

Each punch is acted on by the normal force q. We assume that the pressure and displacement functions near each punch are axisymmetric.

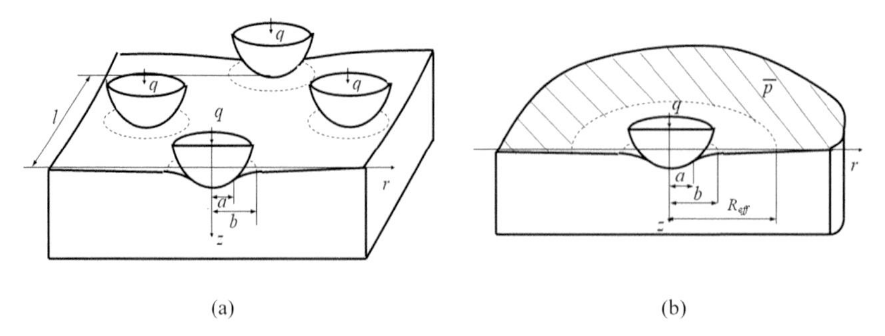

Figure 2.4 Scheme of contact between a periodic system of punches and an elastic half-space (a). Application of the localization method (b).

The value of gap $h(r)$ between the surfaces of the elastic half-space and a punch in the vicinity of each punch is also a function of only the radial coordinate:

$$h(r) = f(r) - f(a) + u_z(r) - u_z(a), \tag{2.11}$$

where $u_z(r)$ is the normal displacement of the surfaces due to their deformation, and a is the radius of contact area.

To take into account adhesion, we assume that the negative pressure $-p_0$ is applied to the surface of the elastic half-space in the ring-shaped region $a \leq r \leq b$ around each contact spot. Two types of adhesion will be considered.

In the case of **molecular adhesion** (Section 2.1.1), we use the Maugis-Dugdale model given by Eqs. (2.6) and (2.7). From Eq. (2.7) for the specific work of adhesion w_a, we obtain the condition to be imposed on the value of the gap at $r = b$:

$$h(b) = \frac{w_a}{p_0}. \tag{2.12}$$

Here the values w_a and p_0 are assumed to be known for a pair of surfaces under consideration.

In the case of **capillary adhesion** (Section 2.1.2), the surface of the elastic half-space is covered by a fluid film of thickness h_1 before the interaction. When the half-space comes into contact with the system of punches, all fluid is gathered into menisci of the same volume around contact spots. From geometrical considerations, the fluid volume in each meniscus is calculated as

$$v = \frac{\sqrt{3}l^2 h_1}{2}. \tag{2.13}$$

In this case, the region $a \leq r \leq b$ is occupied by fluid. The pressure under a curved surface of fluid is lower than the atmospheric pressure by the value p_0 defined by the Laplace formula (2.8). We assume that the atmospheric pressure is zero, the wetting angles for the interacting surfaces θ_1 and θ_2 are zero, and the shape of the punches satisfies the condition $f'(b) << 1$; then for the radii of the menisci, we have $R_1 \approx h(b)/2$ and $R_2 \approx b$.

It is also assumed that $h(b) << b$. As a result, from the Laplace formula (2.8) we come to the following relation for the Laplace capillary pressure:

$$p_0 = \frac{2\gamma_0}{h(b)}. \tag{2.14}$$

By introducing the notation $w_0 = 2\gamma_0$ into Eq. (2.14), we obtain the relation whose form coincides with that of Eq. (2.12), though in this case the value of p_0 is a priori unknown.

Apart from the force F_L associated with the Laplace pressure Eq. (2.9), the surface of the elastic half-space is acted on by the force of surface tension of fluid given by Eq. (2.10). This force acts along the circumference $r = b$ around each punch. Since wetting angles are zero, this force can be represented as $F_s = -2\pi b\gamma_0$. The force F_s is directed tangentially to the surface of the elastic half-space. The following simple estimates show that this force is much smaller than the force F_L due to the Laplace capillary pressure. If the surfaces are not in direct contact, but are separated by the menisci ($h(0)>0$), we have $F_L = -\pi b^2 p_0$, from which it follows that $\frac{F_s}{F_L} = \frac{h(b)}{b} << 1$. When the surfaces of the half-space and the punches are in direct contact ($h(0) = 0$), in accordance with Eq. (2.10), we have $F_L = -\pi\left(b^2 - a^2\right)p_0$, from which we obtain $\frac{F_s}{F_L} = \frac{bh(b)}{b^2 - a^2} \approx f'(b) << 1$. Based on this estimation, below we neglect the force of surface tension of fluid.

Let the volume of fluid in each meniscus be approximately calculated as volume of the gap in the region $a \le r \le b$ around each punch:

$$v_0 = \int_0^{2\pi} \int_a^b rh(r)\,dr\,d\phi. \tag{2.15}$$

2.2.2 Method of solution

We transform the boundary conditions of the problem by using the localization method. In accordance with this method, the stress-strain state near a contact spot is determined by the accurate contact conditions only for the nearest contact spots, the influence of the remaining contact spots being replaced by the action of an averaged pressure (for more detail of the method, see Chapter 1).

We apply the simplest version of the localization method in which only the interaction of one punch with the elastic half-space is taken into account, while the uniform averaged pressure $\overline{p}$ is applied over the region

$r \geq R_{\text{eff}}$ (Fig. 2.4b). The averaged pressure is calculated from the geometrical considerations as

$$\overline{p} = \frac{2q}{\sqrt{3}l^2}. \tag{2.16}$$

The effective radius R_{eff} is determined from the condition that the averaged pressure is equal inside the circle $r \leq R_{\text{eff}}$ and outside it, i.e., $\overline{p} = \frac{q}{\pi R_{\text{eff}}^2}$, from which, taking into account Eq. (2.16), we get

$$R_{\text{eff}} = l\left(\frac{\sqrt{3}}{2\pi}\right)^{1/2}. \tag{2.17}$$

Finally, we have a boundary value problem for the elastic half-space loaded in accordance with the following conditions:

$$\begin{aligned} u_z(r) &= -f(r) - d, & 0 \leq r \leq a \\ p(r) &= -p_0, & a \leq r \leq b \\ p(r) &= 0, & b < r \leq R_{\text{eff}}, \\ p(r) &= \overline{p}, & r > R_{\text{eff}} \end{aligned} \tag{2.18}$$

where d is a constant equal to the distance between the punch top and the unperturbed surface of the elastic half-space.

The normal elastic displacement $u_z(r)$ of the boundary of the half-space is related to the normal pressure $p(r)$ by the following equation valid for the case of axisymmetric loading (Johnson, 1985):

$$u_z(r) = \frac{4}{\pi E^*}\int_0^b p(r')\mathbf{K}\left(\frac{2\sqrt{rr'}}{r+r'}\right)\frac{r'dr'}{r+r'}, \tag{2.19}$$

where $\mathbf{K}(x)$ is the full elliptic integral of the first kind, $E^* = E/(1-\nu^2)$, and E and ν are the Young modulus and Poisson's ratio of the elastic half-space, respectively. The external normal load q acting on each punch is in equilibrium with the contact and adhesion pressures:

$$q = 2\pi\int_0^b rp(r)dr, \tag{2.20}$$

Note that the force q can have both positive and negative values.

Thus, the discrete contact problem taking into account molecular adhesion is specified by Eqs. (2.12), (2.16)–(2.20). The discrete contact problem in the presence of capillary menisci is defined by Eqs. (2.14)–(2.20).

To construct the solution to the problems stated, we represent the function of contact pressure $p(r)$ for $0 \le r \le a$ as $p(r) = p_1(r) - p_0$. Taking into account conditions Eq. (2.18), we reduce Eq. (2.19) to the form

$$-f_1(r) - d_a = \frac{4}{\pi E^*} \int_0^a p_1(r')\mathbf{K}\left(\frac{2\sqrt{rr'}}{r+r'}\right) \frac{r' dr'}{r+r'}, \tag{2.21}$$

where we denote $f_1(r) = f(r) - \frac{4}{\pi E^*}\left[p_0 b \mathbf{E}\left(\frac{r}{b}\right) + \bar{p} R \mathbf{E}\left(\frac{r}{R_{\text{eff}}}\right)\right]$ and $d_a = d + \frac{4}{\pi E^*}\bar{p} \int_0^\infty \mathbf{K}\left(\frac{2\sqrt{rr'}}{r+r'}\right) \frac{r' dr'}{r+r'}$. Here, $\mathbf{E}(x)$ is the full elliptic integral of the second kind.

Note that the distance d is infinite since the half-space is loaded over the infinite region. But the constant d_a is a finite value specifying the distance between the punch top and the boundary of the half-space uniformly loaded by the averaged pressure $\bar{p}$. The value d_a will be called the additional displacement of the punches.

Eq. (2.21) can be considered an equation for the contact pressure $p_1(r)$ under a smooth punch whose shape is described by the function $f_1(r)$ with no additional loads applied on the half-space. We solve this problem by using the Galin solution for an arbitrary axisymmetric punch (Galin and Gladwell, 2008) expanding integrands into series with subsequent term-to-term integrating them (Goryacheva and Makhovskaya, 1999). As a result, the contact pressure of the original problem is obtained in the form

$$p(r) = \frac{AE^* a^{2n-1}}{\pi}\left[\frac{(2n)!!}{(2n-1)!!}\right]^2 \sqrt{1-\frac{r^2}{a^2}} \sum_{k=1}^{n} \frac{(2k-3)!!}{(2k-2)!!} \frac{r^{2(n-k)}}{a^{2(n-k)}}$$
$$- p_0 \operatorname{arccot}\sqrt{\frac{a^2-r^2}{b^2-a^2}} + \frac{2}{\pi}\bar{p} \arctan\sqrt{\frac{a^2-r^2}{R_{\text{eff}}^2-a^2}}, \quad r \le a. \tag{2.22}$$

We also obtain the following expressions for the additional displacement of the punches

$$d_a = -\frac{(2n)!!}{(2n-1)!!}Aa^{2n} + \frac{2p_0 b}{E^*}\sqrt{1-\frac{a^2}{b^2}} - \frac{2\overline{p}R_{\text{eff}}}{E^*}\sqrt{1-\frac{a^2}{R_{\text{eff}}^2}} \tag{2.23}$$

and for the normal load applied to each punch

$$q = \frac{\frac{(2n)!!}{(2n+1)!!}2\pi E^* Ana^{2n+1} - \pi p_0 b^2\left(\arccos\frac{a}{b} + \frac{a}{b}\sqrt{1-\frac{a^2}{b^2}}\right)}{\arccos\frac{a}{R_{\text{eff}}} + \frac{a}{R_{\text{eff}}}\sqrt{1-\frac{a^2}{R_{\text{eff}}^2}}}. \tag{2.24}$$

The gap between the contacting surfaces is obtained in the form

$$\begin{aligned} h(r) &= \frac{2}{\pi}\arccos\frac{a}{r}\left(Ar^{2n} - \frac{(2n)!!\,Aa^{2n}}{(2n-1)!!} - \frac{2p_0 b}{E^*}\sqrt{1-\frac{a^2}{b^2}}\right) \\ &+ \frac{2Ar^{2n}}{\pi}\sqrt{\frac{r^2}{a^2}-1}\sum_{k=1}^{n}\frac{(2k-2)!!}{(2k-1)!!}\left(\frac{a}{r}\right)^{2k} \\ &- \frac{4}{\pi E^*}\left\{p_0 b\left[\mathbf{E}\left(\frac{r}{b}\right) - \mathbf{E}\left(\frac{a}{r},\frac{r}{b}\right)\right] + \overline{p}R_{\text{eff}}\left[\mathbf{E}\left(\frac{r}{R_{\text{eff}}}\right) - \mathbf{E}\left(\frac{a}{R_{\text{eff}}},\frac{R_{\text{eff}}}{b}\right)\right]\right\}. \end{aligned} \tag{2.25}$$

Taking into account Eq. (2.12) for the maximum value of the gap, from Eq. (2.25) at $r = b$, we obtain the condition that together with Eq. (2.24) serves to determine the unknown values a and b in the problem of molecular adhesion.

For the case of capillary adhesion in which we have one more unknown value—the Laplace pressure p_0—we substitute Eq. (2.25) for the gap into Eq. (2.15) for the fluid volume. As a result, the following expression is obtained for the fluid volume in each meniscus:

$$v_0 = 2\,Aa^{2n+1}\sqrt{\frac{b^2}{a^2}-1}\left[\frac{(2n)!!(2n-1)}{(2n+1)!!}+\frac{1}{n+1}\sum_{k=1}^{n}\frac{(2k)!!}{(2k+1)!!}\left(\frac{b}{a}\right)^{2(n-k)}\right]$$
$$+\,2\,Aa^{2n-1}b^2\arccos\frac{a}{b}\left[\frac{(2n)!!}{(2n-1)!!}+\frac{b^{2n}}{(n+1)a^{2n}}\right]$$
$$-\,\frac{4p_0b^3}{3E^*}\left[4-\frac{3a}{b}-\frac{a^3}{b^3}-3\arccos\frac{a}{b}\sqrt{1-\frac{a^2}{b^2}}\right]$$
$$-\,\frac{4\bar{p}R_{\text{eff}}}{E^*}\left[\left(R_{\text{eff}}^2-a^2\right)\mathbf{K}\left(\frac{a}{R_{\text{eff}}}\right)-\left(R_{\text{eff}}^2+a^2\right)\mathbf{E}\left(\frac{a}{R_{\text{eff}}}\right)-\left(R_{\text{eff}}^2-b^2\right)\mathbf{K}\left(\frac{b}{R_{\text{eff}}}\right)\right]$$
$$-\,\frac{4\bar{p}R_{\text{eff}}}{E^*}\left[\left(R_{\text{eff}}^2+b^2\right)\mathbf{E}\left(\frac{b}{R_{\text{eff}}}\right)-(b^2-a^2)\mathbf{K}\left(\frac{a}{b}\right)+(b^2+a^2)\mathbf{E}\left(\frac{a}{b}\right)-3ab+\frac{a^3}{b}\right]. \tag{2.26}$$

The solution is significantly simpler for the case where no direct contact occurs between the punches and the half-space. In the case of molecular adhesion, for the load and additional displacement, we have:

$$q=-\pi p_0b^2,$$
$$d_a=-Ab^{2n}+\frac{4p_0b}{\pi E^*}\left[1-\frac{b}{R_{\text{eff}}}\mathbf{E}\left(\frac{b}{R_{\text{eff}}}\right)\right]+\frac{\gamma}{p_0}. \tag{2.27}$$

Eq. (2.27) should be amplified by the condition of constant fluid volume, which follows from Eq. (2.15):

$$v_0=\frac{8p_0b^2R_{\text{eff}}}{3E^*}\left[E\left(\frac{b}{R_{\text{eff}}}\right)\left(1+\frac{b^2}{R_{\text{eff}}^2}\right)-K\left(\frac{b}{R_{\text{eff}}}\right)\left(1-\frac{b^2}{R_{\text{eff}}^2}\right)-2\frac{b}{R_{\text{eff}}}\right]$$
$$+\,\pi b^2\left(\frac{Ab^{2n}}{n+1}+d_a\right). \tag{2.28}$$

Thus the complete system of equation to calculate the contact pressure distribution at each contact spot and gap between the surfaces, taking into account the density of contact spots, consists of Eqs. (2.12), (2.22)–(2.25), and (2.27) for dry contacting surfaces and of Eqs. (2.14), (2.22)–(2.28) for wet surfaces.

2.3 Effect of molecular adhesion in normal discrete contact

In this section, numerical results are presented for an elastic half-space indented by a periodic system of punches in the presence of molecular adhesion. This contact configuration can describe a regular microrelief applied on a surface.

The results presented are obtained by numerical analysis of relations Eqs. (2.12), (2.22)−(2.25) for contact and Eq. (2.27) for contactless interaction. The following dimensionless values are calculated: the contact pressure $P = p/\pi E^*$, normal displacement of the elastic half-space surface $U_z = u_z/D$ ($D = A^{-1/(2n-1)}$ is the characteristic size of an asperity), normal load applied to each punch $Q = q/\pi E^* D^2$, additional displacement of the punches $\delta_a = d_a/D$, radius of the contact region $\alpha = a/D$, and outer radius of the adhesion region $\beta = b/D$.

Model parameters defining the process of molecular adhesive interaction between an elastic half-space and regular system of punches are the following. The first group of the parameters are $K = w_a/2\pi E^* D$ and $P_0 = p_0/\pi E^*$, which depend on the characteristics w_a and p_0 of the Maugis-Dugdale model of adhesion given by Eqs. (2.6) and (2.7), elastic properties of the half-space, and size of the punches. The second group includes the parameters n and $L = R_{\mathrm{eff}}/D = 3^{1/4}l/\sqrt{2\pi}D$, which describe the shape of the punches and spacing between them, respectively. The results obtained are compared to the case of the adhesive interaction of a single punch with an elastic half-space ($L \to \infty$).

In Fig. 2.5, the distributions of the dimensionless contact pressure P over the dimensionless coordinate r/D are presented for $K = 5 \times 10^{-5}$, $P_0 = 2 \times 10^{-2}$, and the same normal load $Q = 2 \times 10^{-3}$ for two different shapes of the punch, $n = 1$ and $n = 2$. Thin lines correspond to the dimensionless spacing of punches $L = 0.15$. Thick lines are constructed for a single punch ($L \to \infty$). The results indicate that taking into account the influence of other punches leads to a decrease in the contact radius α and adhesion radius β and to an increase in the maximum contact pressure.

Fig. 2.6 depicts the value $U_z(1) - U_z(\rho)$ as a function of $\rho = r/a$, which illustrates the shape of the surface of the elastic half-space outside the contact region. The values of the parameters K, P_0, and Q are the same as

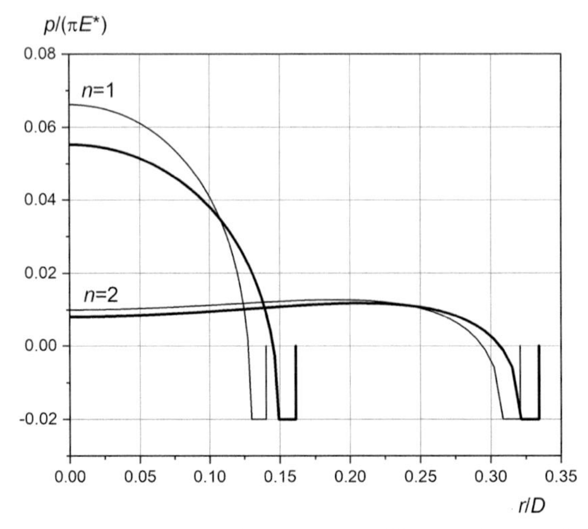

Figure 2.5 Contact pressure distributions for two different shapes of the punches for the case of molecular adhesion.

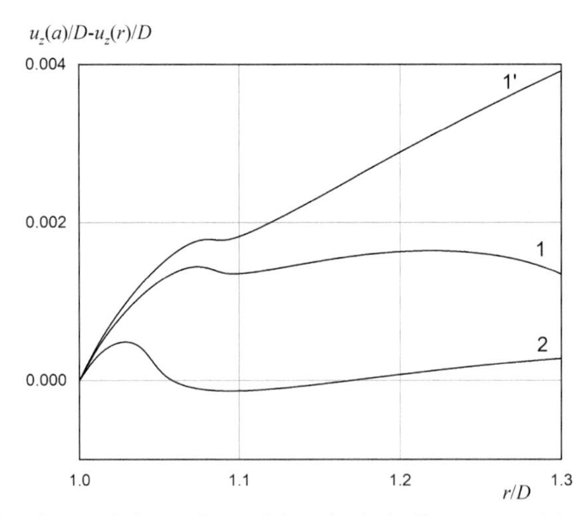

Figure 2.6 The shape of the surface of the elastic half-space outside the contact region in the case of molecular adhesion.

those for Fig. 2.5. The results are obtained for $n = 1$ (curves 1, 1′) and $n = 2$ (curve 2). Curve 1 corresponds to $L = 0.2$, while curves 1′ and 2 to $L = 0.5$. It is apparent that as the spacing of punches decreases, the deflection of the elastic half-space boundary outside the contact area increases.

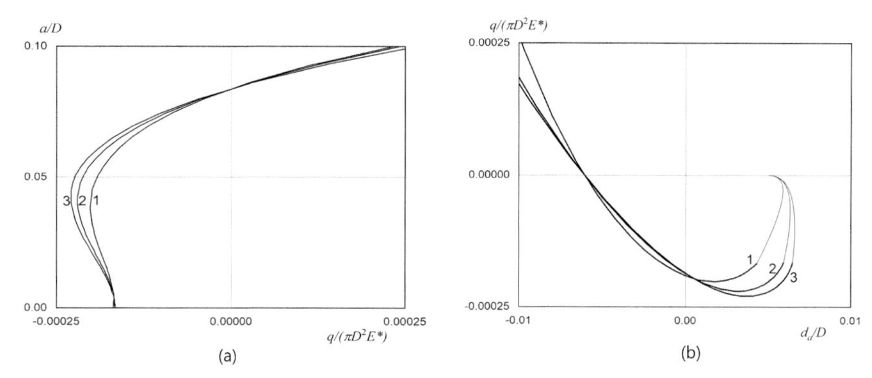

Figure 2.7 Dimensionless contact radius versus normal load (a) and load versus additional displacement (b) for the case of molecular adhesion.

The contact characteristics calculated for $K = 5 \times 10^{-5}$, $P_0 = 2 \times 10^{-2}$, $n = 1$ are presented in Fig. 2.7 for two values of the dimensionless spacings of the punches $L = 0.15$ and $L = 0.60$ (curves 1 and 2, respectively). Curves 3 correspond to the case of the single punch ($L \to \infty$). The graphs of the dimensionless contact radius α versus the normal load Q presented in Fig. 2.7a show that for positive loads, reducing the spacing of punches leads to a decrease in the contact radius, as it is the case for the discrete contact without adhesion (see Chapter 1). However, in the case of negative loads, reducing the spacing of punches leads to an increase in the contact radius α.

In Fig. 2.7b, the normal load Q is presented as a function of the additional displacement δ_a. Thick lines correspond to direct contact between the surfaces and are calculated by Eqs. (2.12), (2.22)–(2.25). Thin lines correspond to the contactless case where the surfaces interact only via adhesive pressure and are calculated by Eq. (2.27). The results indicate that as the spacing of punches L decreases, the displacement δ_a decreases for $Q > 0$ and increases for $Q < 0$.

The analysis of the influence of the punch shape, which is defined by the parameter n, allows us to conclude that this parameter has a significant effect on the contact pressure distribution, contact radius, and other contact characteristics in the case of molecular adhesion.

The analysis of the effect of mutual influence of the punches through the elastic half-space makes it possible to establish that the mutual influence of neighboring punches becomes noticeable only for relatively small distance between them ($L < 0.5$). Reducing the spacing of punches has different effects depending on the sign of the applied normal load:

- If the force applied to each punch is positive (pressing the punches to the elastic half-space), then reducing the spacing of punches leads to an increase in the contact pressure, decrease in the contact area and area of adhesion, as well as to a decrease in the additional displacement.
- If the force is negative (pulling the punches apart from the half-space), then reducing the spacing of punches leads to a decrease in the contact pressure and an increase of the areas of contact and adhesion and in the additional displacement.

The external normal force applied to the punches is a nonmonotone and ambiguous function of the displacement, which is characteristic for the adhesion interaction of elastic bodies (Goryacheva and Makhovskaya, 2001, 2008).

2.4 Effect of capillary adhesion in normal discrete contact

In this section, the results are presented for the case of capillary adhesion between a periodic system of punches and an elastic half-space. The results presented are obtained by numerical analysis of Eqs. (2.14), (2.22)−(2.26) for the case of contact and of Eqs. (2.27)−(2.28) for the contactless case. In this case, the model parameters are $K = \gamma_0 / \pi E^* D$, characterizing the surface tension of the fluid, and the elastic modulus of the half-space, $H_1 = h_1 / D$, which is the dimensionless fluid film thickness covering the half-space before the interaction, and the parameters n and L describing the shape of the punches and spacing between them, respectively.

The results of the calculations performed show that reducing the spacing of punches, which implies reducing the fluid volume in each meniscus, leads to a sharp increase in the absolute value of the Laplace capillary pressure and decrease in the width $\beta - \alpha$ of the region occupied by fluid (Makhovskaya, 2003). However, the shape of the contact pressure distribution does not significantly change as the spacing of punches is changing.

The analysis of the results indicates also that the contact radius α non-monotonically depends on the spacing of punches L. This can be seen on the graphs of α as a function of L that are presented in Fig. 2.8a. These results are calculated for $n = 1$, $K = 5 \times 10^{-5}$, and $Q = 10^{-3}$. Different curves correspond to different thickness of fluid film covering the surface

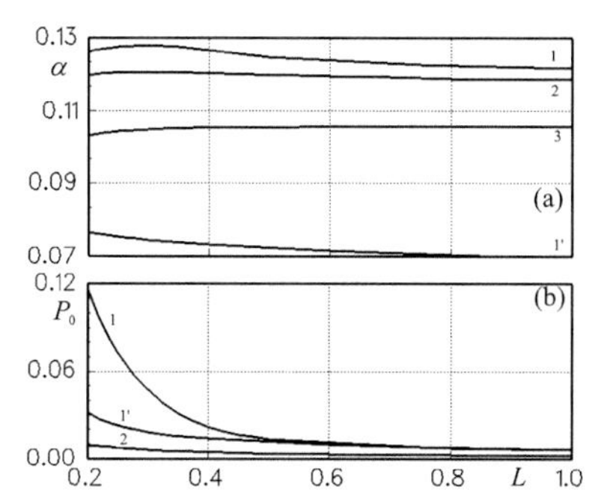

Figure 2.8 Dimensionless contact radius (a) and Laplace pressure (b) as functions of the spacing of punches for the case of capillary adhesion.

before the interaction: $H_1 = 10^{-4}$ (curve 1) and $H_1 = 10^{-3}$ (curve 2). Curve 3 corresponds to the case of no fluid and no adhesion ($H_1 = 0$). Curve 1′ is constructed for $H_1 = 10^{-4}$ and negative load $Q = -10^4$.

The specifics of the results presented in Fig. 2.8a can be explained by the effect of the following two mechanisms as the spacing of punches changes:

- The first mechanism is due to mutual influence of the punches through the elastic half-space that takes place only for relatively small values of L. For $Q > 0$, this leads to a decrease in the contact radius α as the spacing L decreases. For $Q < 0$ the mutual influence leads to increasing α as L decreases.
- The second mechanism is associated with a prescribed thickness of the original liquid film H_1 and the total volume of fluid being held constant during the interaction, due to which as the punches come closer to each other, the volume of each meniscus decreases. As a result of this, the contact radius α increases as the spacing L decreases, irrespective of the sign of the load Q.

For $Q > 0$, these two mechanisms have opposite effects on the contact radius that is responsible for the nonmonotonic character of curves 1 and 2 in Fig. 2.8a. For $Q < 0$, these two mechanisms amplify each other, which results in a monotonic graph (curve 1′).

The graphs of the absolute value of the dimensionless Laplace pressure P_0 versus the spacing of punches L are presented in Fig. 2.8b. Curves 1, 2, and 1′ correspond to the same values of the parameters as those in Fig. 2.8a. It is established that for any values of the normal load, the absolute value of the Laplace pressure grows when reducing the spacing of punches L. The results indicate also that the outer radius of the menisci β monotonically

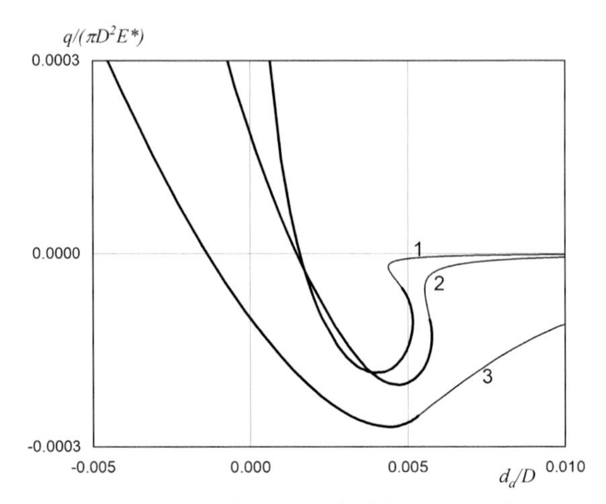

Figure 2.9 Dimensionless load as a function of additional displacement for the case of capillary adhesion.

decreases when reducing L for any values of the load Q. The last two results are accounted for by decreasing the volume of fluid in each meniscus as the punches move closer to each other.

In Fig. 2.9, the dimensionless normal load Q versus the additional displacement of the punches is presented for n $= 1$, $K = 5 \times 10^{-5}$, $H_1 = 10^{-5}$, and various values of the dimensional spacing of punches: $L = 0.2$, 0.4, 2 (curves 1, 2, 3, respectively). Thick lines correspond to the case of direct contact between the half-space and the punches, and thin lines correspond to the surfaces separated by the fluid bridges. The results indicate that (as it was the case for molecular adhesion) the graphs of the load versus displacement are ambiguous but not for all values of the parameters. In particular, as the spacing of punches decreases, the region of ambiguity of the curves becomes wider.

As the results of analysis indicate, the shape of the punches defined by the number n has a considerable influence on the shape and absolute values of the contact pressure distribution, as it was for molecular adhesion. The effect of spacing of punches on the contact characteristics for capillary adhesion is different from that of molecular adhesion:

- If the normal load is negative, the contact area increases as the spacing of punches decreases.
- If the normal load is positive, the contact area nonmonotonically depends on the spacing of punches.

Also, it is established that as the spacing of punches decreases, the size of menisci also decreases while the Laplace pressure increases, irrespective of the sign of the applied normal load. The normal load as a function of the additional displacement is nonmonotone and ambiguous.

The role of surface tension of fluid defined by the parameter K in the conditions of discrete contact is similar to that for a single contact (Goryacheva and Makhovskaya, 1999; Makhovskaya and Goryacheva, 1999). As K increases, the contact radius and the outer radius of menisci also increases. The normal load as a function of the additional displacement becomes ambiguous for high enough values of the parameter K.

It should be stressed that the results presented above correspond to a fixed load Q applied to each punch. This formulation made it possible to analyze the pure effect of mutual influence of the punches through the elastic half-space.

2.5 Conclusion

The solutions of the problems of contact and contactless interaction between an elastic half-space and a periodic system of axisymmetric asperities are constructed for the cases of adhesion of a different nature — molecular and capillary.

The solutions obtained made it possible to analyze the effect of the molecular and capillary adhesion parameters (specific work of adhesion of the surfaces, surface tension of the fluid, volume of fluid in menisci, etc.) and the microgeometry parameters (shape of the contacting surface of an asperity, density of arrangement of asperities on the surface) on the contact characteristics, such as real contact pressures, areas of contact and adhesion, load-distance curve, when the surfaces approach and retract from each other.

The results obtained in this chapter allow one to develop methods of control of the contact characteristics by choosing an optimal surface relief with taking into account physical and mechanical properties of surface layers and fluid films covering them.

References

Adams, G.G., 2004. Adhesion at the wavy contact interface between two elastic bodies. J. Appl. Mech. Transac. ASME. 71 (6), 851–856. https://doi.org/10.1115/1.1794702.

Barber, J.R., Ciavarella, M., 2014. JKR solution for an anisotropic half space. J. Mech. Phys. Solid. 64 (1), 367–376. https://doi.org/10.1016/j.jmps.2013.12.002.

Borodich, F.M., Galanov, B.A., Keer, L.M., Suárez-Álvarez, M., 2014a. The JKR-type adhesive contact problems for transversely isotropic elastic solids. Mech. Mater. 75, 34–44. https://doi.org/10.1016/j.mechmat.2014.03.011.
Borodich, F.M., Galanov, B.A., Suarez-Alvarez, M.M., 2014b. The JKR-type adhesive contact problems for power-law shaped axisymmetric punches. J. Mech. Phys. Solid. 68 (1), 14–32. https://doi.org/10.1016/j.jmps.2014.03.003.
Briggs, G.A.D., Briscoe, B.J., 1977. The effect of surface topography on the adhesion of elastic solids. J. Phys. Appl. Phys. 10 (18), 2453–2466. https://doi.org/10.1088/0022-3727/10/18/010.
Butt, H.J., Barnes, W.J.P., del Campo, A., Kappl, M., Schönfeld, F., 2010. Capillary forces between soft, elastic spheres. Soft Matter 6, 5930–5936. https://doi.org/10.1039/c0sm00455c.
Derjaguin, B., 1934. Untersuchungen über die Reibung und Adhäsion, IV - Theorie des Anhaftens kleiner Teilchen. Kolloid Z. 69 (2), 155–164. https://doi.org/10.1007/BF01433225.
Derjaguin, B.V., Muller, V.M., Toporov, Y.P., 1975. Effect of contact deformations on the adhesion of particles. J. Colloid Interface Sci. 53 (2), 314–326. https://doi.org/10.1016/0021-9797(75)90018-1.
Fan, H., Gao, Y.X., 2001. Elastic solution for liquid-bridging- induced microscale contact. J. Appl. Phys. 90 (12), 5904–5910.
Galin, L.A., Gladwell, G.M.L. (Eds.), 2008. Contact Problems.
Gao, C., Tian, X., Bhushan, B., 1995. A meniscus model for optimization of texturing and liquid lubrication of magnetic thin film rigid disks. Tribol. Trans. 38, 201–212.
Goryacheva, I.G., Makhovskaya, Y.Y., 1999. Capillary adhesion in the contact between elastic solids. J. Appl. Math. Mech. 63 (1), 117–125. https://doi.org/10.1016/S0021-8928(99)00017-9.
Goryacheva, I.G., Makhovskaya, Y.Y., 2001. Adhesive interaction of elastic bodies. J. Appl. Math. Mech. 65 (2), 273–282. https://doi.org/10.1016/S0021-8928(01)00031-4.
Goryacheva, I.G., Makhovskaya, Y.Y., 2004. An approach to solving the problems on interaction between elastic bodies in the presence of adhesion. Dokl. Phys. 49 (9), 534–538. https://doi.org/10.1134/1.1810581.
Goryacheva, I., Makhovskaya, Y., 2008. Adhesion effects in contact interaction of solids. Compt. Rendus Mec. 336 (1–2), 118–125. https://doi.org/10.1016/j.crme.2007.11.003.
Greenwood, J.A., 1997. Adhesion of elastic spheres. Proc. Math. Phys. Eng. Sci. 453 (1961), 1277–1297. https://doi.org/10.1098/rspa.1997.0070.
Hughes, B.D., White, L.R., 1979. 'Soft' contact problems in linear elasticity. Q. J. Mech. Appl. Math. 32 (4), 445–471. https://doi.org/10.1093/qjmam/32.4.445.
Hui, C.Y., Lin, Y.Y., Baney, J.M., Kramer, E.J., 2001. The mechanics of contact and adhesion of periodically rough surfaces. J. Polym. Sci. B Polym. Phys. 39 (11), 1195–1214. https://doi.org/10.1002/polb.1094.
Israelachvili, J., 1992. Intermolecular and Surface Forces. Academic Press, London.
Jin, F., Guo, X., Wan, Q., 2016. Revisiting the Maugis-Dugdale adhesion model of elastic periodic wavy surfaces. J. Appl. Mech. Trans. ASME 83 (10). https://doi.org/10.1115/1.4034119.
Johnson, K., 1985. Contact Mechanics. Cambridge University Press.
Johnson, K.L., 1995. The adhesion of two elastic bodies with slightly wavy surfaces. Int. J. Solid Struct. 32 (3–4), 423–430. https://doi.org/10.1016/0020-7683(94)00111-9.
Johnson, K.L., Greenwood, J.A., 1997. An adhesion map for the contact of elastic spheres. J. Colloid Interface Sci. 192 (2), 326–333. https://doi.org/10.1006/jcis.1997.4984.
Johnson, K.L., Kendall, K., Roberts, A.D., 1971. Surface energy and the contact of elastic solids. Proc. R Soc. London. Series A. 324 (1558), 301–313.

Makhovskaya, Y.Y., 2016. Modeling contact of indenter with elastic half-space with adhesive attraction assigned in arbitrary form. J. Frict. Wear 37 (4), 301–307. https://doi.org/10.3103/S1068366616040103.
Makhovskaya, Y.Y., Goryacheva, I.G., 1999. Combined effect of capillarity and elasticity in contact interaction. Tribol. Int. 32 (9), 507–515. https://doi.org/10.1016/S0301-679X(99)00080-8.
Makhovskaya, Y.Y., 2003. Discrete contact of elastic bodies in the presence of adhesion. Izv. RAS Mech. Solid 2, 49–60.
Maugis, D., 1992. Adhesion of spheres: the JKR-DMT transition using a dugdale model. J. Colloid Interface Sci. 150 (1), 243–269. https://doi.org/10.1016/0021-9797(92)90285-T.
Maugis, D., 2000. Contact, Adhesion and Rupture of Elastic Solids. Springer.
Maugis, D., Gauthier-Manuel, B., 1994. JKR-DMT transition in the presence of a liquid meniscus. J. Adhes. Sci. Technol. 8 (11), 1311–1322. https://doi.org/10.1163/156856194X00627.
Megias-Alguacil, D., Gauckler, L.J., 2009. Capillary forces between two solid spheres linked by a concave liquid bridge: regions of existence and forces mapping. AIChE J. 55 (5), 1103–1109. https://doi.org/10.1002/aic.11726.
Payam, A.F., Fathipour, M., 2011. A capillary force model for interactions between two spheres. Particuology 9 (4), 381–386. https://doi.org/10.1016/j.partic.2010.11.004.
Purtov, J., Gorb, E.V., Steinhart, M., Gorb, S.N., 2013. Measuring of the hardly measurable: adhesion properties of anti-adhesive surfaces. Appl. Phys. Mater. Sci. Process 111 (1), 183–189. https://doi.org/10.1007/s00339-012-7520-3.
Rabinovich, Y.I., Esayanur, M.S., Moudgil, B.M., 2005. Capillary forces between two spheres with a fixed volume liquid bridge: theory and experiment. Langmuir 21 (24), 10992–10997. https://doi.org/10.1021/la0517639.
Rabinowicz, E., 1965. Friction and Wear of Materials. Wiley, New York.
Soldatenkov, I.A., 2012. The use of the method of successive approximations to calculate an elastic contact in the presence of molecular adhesion. J. Appl. Math. Mech. 76 (5), 597–603. https://doi.org/10.1016/j.jappmathmech.2012.11.005.
Soldatenkov, I.A., 2013. The contact problem with the bulk application of intermolecular interaction forces (a refined formulation). J. Appl. Math. Mech. 77 (6), 629–641. https://doi.org/10.1016/j.jappmathmech.2014.03.007.
Soldatenkov, I.A., 2016. The contact problem with the bulk application of intermolecular interaction forces: the influence function for an elastic 'layer–half-space' system. J. Appl. Math. Mech. 80 (4), 351–358. https://doi.org/10.1016/j.jappmathmech.2016.09.011.
Tabor, D., 1977. Surface forces and surface interactions. J. Colloid Interface Sci. 58 (1), 2–13. https://doi.org/10.1016/0021-9797(77)90366-6.
Wei, Z., Zhao, Y.P., 2004. Adhesion elastic contact and hysteresis effect. Chin. Phys. 13 (8), 1320–1325. https://doi.org/10.1088/1009-1963/13/8/024.
Wu, J.J., 2012. Numerical simulation of the adhesive contact between a slightly wavy surface and a half-space. J. Adhes. Sci. Technol. 26 (1–3), 331–351. https://doi.org/10.1163/016942411X576527.
Zakerin, M., Kappl, M., Backus, E.H.G., Butt, H.-J., Schönfeld, F., 2013. Capillary forces between rigid spheres and elastic supports: the role of young's modulus and equilibrium vapor adsorption. Soft Matter 9 (17), 4534–4543. https://doi.org/10.1039/c3sm27952a.
Zhijun, Z., Jilin, Y., 2007. Using the Dugdale approximation to match a specific interaction in the adhesive contact of elastic objects. J. Colloid Interface Sci. 310 (1), 27–34. https://doi.org/10.1016/j.jcis.2007.01.042.

CHAPTER 3

Additional displacement due to microgeometry of contacting bodies

Much experimental research of machine and instrument parts contact stiffness has shown a difference between experiment and calculations performed for smooth elastic bodies. To remove this discrepancy, contact problems for elastic bodies taking into account their surface microgeometry must be formulated and solved. The first formulation of such a problem was done by Shtaierman (1949). Further research interest was related to both experimental determination of influence of roughness parameters on contact characteristics (Kragelsky, 1965; Kragelsky et al., 1982) and formulation of new contact problems for bodies with various macro- and microgeometry by using specified boundary conditions (Alexandrov and Kudish, 1979; Galanov, 1984, 1985; Goryacheva, 1979, 2006; Rabinovich, 1975; Teply, 1981).

Influence of a rough layer on the contact interaction of elastic bodies can be quantitatively estimated on the base of discrete contact mechanics. Here, the two basic approaches—statistical and deterministic—can be applied. Statistical approach assumes interaction of two single asperities of regular shape, the number of contacting asperities depending on probability of their heights. Displacements of asperities are summarized with displacements of the whole curvilinear elastic body. Then the contact problem is solved with neglecting, as a rule, the elastic interaction due to mutual influence of asperities. One of the first studies in this direction was published by Greenwood and Tripp (1967). The comparative review on statistical approach was given by Sviridenok et al. (1990).

In this chapter, the deterministic approach for evaluation of the additional displacement due to a surface roughness in contact problems is used, which allows us to take into account the elastic interaction between asperities. Based on this approach, the additional displacement as a function of nominal pressure is analyzed for various types of surface microgeometry.

Discrete Contact Mechanics with Applications in Tribology
ISBN 978-0-12-821799-3
https://doi.org/10.1016/B978-0-12-821799-3.00001-7

3.1 Additional displacement function (2-D analysis)

Additional displacement function describing the compliance of the surface roughness was firstly introduced by Shtaierman (1949). He assumed that in penetration of a rough body of specified macrogeometry $f(x)$ into the elastic half-plane, the elastic displacements at each point of nominal contact were defined by the sum of the displacements of a half-plane from the nominal (average) pressure $p(x)$ and the displacements related to local deformation of roughness. Shtaierman (1949) assumed that this additional displacement is proportional to the nominal pressure, i.e., $u_{z2} = kp(x)$, where k is a coefficient which depends on the surface microgeometry. By satisfying the contact condition, the following integral equation has been deduced for the nominal contact pressure $p(x)$ (Shtaierman, 1949):

$$-\frac{\partial f(x)}{\partial x} = k\frac{\partial p(x)}{\partial x} + \frac{2(1-\nu^2)}{\pi E}\int_{-a}^{a}\frac{p(\xi)}{x-\xi}\mathrm{d}\xi. \tag{3.1}$$

Schtaierman proposed to determine the coefficient k in this equation by experiment. Attempts to measure the additional displacement function experimentally were performed by Kragelsky (1965). The experiments were carried out on a developed test bench for determining contact separation of circular specimens at first and successive loading. The bench design allowed avoiding the influence of specimen bulk deformation and determining only the displacements of surface asperities. However, a general methodic of experimental determination of the additional displacement function has not been developed yet.

Note that the additional displacement due to surface roughness can be calculated analytically based on the solutions of the contact problems for deformable bodies with regular microgeometry.

For a rigid body with wavy surface penetrating into the elastic half-plane, it can be calculated based on the solutions of periodic contact problems in 2-D formulation. Under the assumption that there is no friction in the contact zones $(-a, a)$, the elastic displacements in contact with a wavy surface are determined from the following expression (Shtaierman, 1949):

$$u(x) = \frac{2(1-\nu^2)}{\pi E}\int_{-a}^{a} p(t)\ln\left|2\sin\frac{\pi(t-x)}{l}\right|dt + C \tag{3.2}$$

Here l is a period of the waviness, and C is an indefinite constant.

Let us assume that the wavy surface described by the periodic function $F(x)$ penetrates into the elastic half-plane under the nominal pressure $\overline{p}$. Its value is determined by the ratio of the linear load P per a single period to the value l of the period: $\overline{p} = P/l$. It follows from the equilibrium equation that

$$P = \int_{-a}^{a} p(x)dx. \tag{3.3}$$

So, the nominal pressure $\overline{p}$ corresponding to the contact zone half width a is given by

$$\overline{p}(a) = \frac{1}{l}\int_{-a}^{a} p(x)dx. \tag{3.4}$$

The pressure distribution $p(x)$ is determined from Eq. (3.2) if we take into account the contact condition

$$u(x) = F(x) + C_1. \tag{3.5}$$

According to the theory given in Chapter 1, the additional displacement for the wavy surface is specified by the following expression:

$$C[\overline{p}] = -F(x_0) + \frac{2(1-\nu^2)}{\pi E} \left[\int_{-l/2}^{l/2} p(\xi)\ln|x_0 - \xi|d\xi - \overline{p}\int_{-l/2}^{l/2} \ln|x_0 - \xi|d\xi\right] + C_2. \tag{3.6}$$

Here, the smooth function $F(x)$ describes the shape of the surface regular relief, and the constant C_2 depends on a selected displacement datum point. If the surface relief has a periodic distribution of asperities heights, and the highest point is located in the origin of coordinates, then the additional displacements should be defined at the point $x_0 = 0$. Thus, the additional displacement function can be written in the following form:

$$C[\overline{p}] = -(\overline{u}_z(0) - \overline{u}_{z\infty}(0)), \tag{3.7}$$

where $\overline{u}_z(0)$ is the displacement of an asperity peak under the pressure distribution $p(x)$, which acts on the period l and is determined from Eqs. (3.2) and (3.5), and $\overline{u}_{z\infty}(0)$ is the displacement of the asperity peak under the nominal pressure.

For numerical calculation of the function Eq. (3.6) the following additional condition, which is implemented with numerical integration of Eqs. (3.6) and (3.7), is used:

$$\lim_{x\to\infty}\left(\overline{u}_z(x)-\overline{u}_{z\infty}(x)\right)=0. \tag{3.8}$$

It follows from Eq. (3.6), that the additional displacement function is considerably influenced by the contact pressure distribution $p(x)$ on each period. The pressure distribution, in turn, is determined by the shape $F(x)$ of the wavy surface, as it follows from Eqs. (3.2) and (3.5).

The additional displacement function in contact of the wavy surface described by the function $F(x) = \Delta \sin lx$ was calculated by Kuznetsov (1978):

$$C(\overline{p})=\frac{\Delta\overline{p}}{p^*}\left(1-\ln\frac{\overline{p}}{p^*}\right). \tag{3.9}$$

Here, $p^* = \pi E\Delta/(l(1-\nu^2))$, and E and ν are the elastic modulus and Poisson ratio of the half-space.

The close-form solution of the plane contact problem was found by Tsukanov (2017) for the waviness, defined by the following function:

$$f(x,m)=\Delta\left[1-\frac{(m+1)\cos(2\pi x/l)}{|m\cos(2\pi x/l)|+1}\right], \tag{3.10}$$

where m is a shape parameter. At $m=0$, Eq. (3.10) becomes a simple sinusoid: $f(x) = \Delta(1 - \cos(2\pi x/L))$. The radii of curvature of the profile peak and valley increase with growth of m (Fig. 3.1).

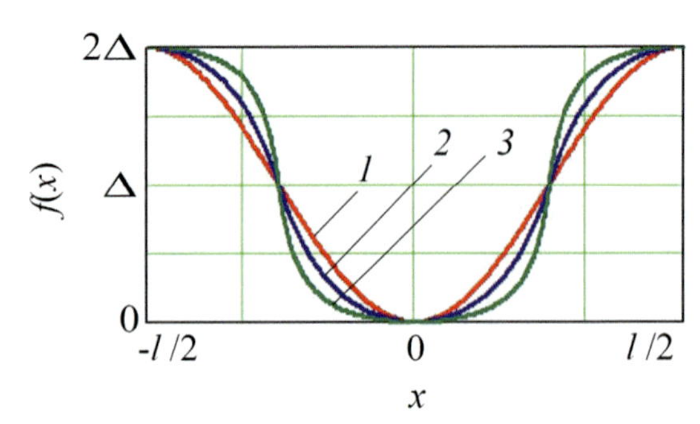

Figure 3.1 Waviness profiles on a single period at different values of m: (1) $m = 0$; (2) $m = 1$; (3) $m = 4$.

The normal pressure in contact of the wavy surface, described by Eq. (3.10), and elastic half-space is determined by the function (Tsukanov, 2017):

$$p(x,m) = -J(x,m)\frac{\sqrt{2}\pi\Delta(1-\nu^2)}{lE}|\cos(\pi x/l)| \times \sqrt{\cos(2\pi x/l) - \cos(2\pi a/l)}, \quad (3.11)$$

where $J(x, m)$ is the following function:

$$J(x,m) = (m+1)^2(|m\cos(2\pi x/l)|+1)^{-2}(m\cos(2\pi a/l)+1)^{-1}. \quad (3.12)$$

Substitution of Eq. (3.11) into Eq. (3.6) and calculation of the integrals let us determine the additional displacement function for the surface waviness described by the periodic function Eq. (3.10).

Graphs of the dimensionless additional displacement function for the waviness characterized by various values of m at $\Delta = 1$ mm and $l = 10$ mm, as a function of the dimensionless nominal pressure $\overline{p}/p^*$, where $p^* = -\pi\Delta E^*/l$, are presented in Fig. 3.2.

The results indicate that with an increase of the parameter m, the additional displacement decreases (Goryacheva and Tsukanov, 2018). It should also be noted that the additional displacement increases linearly with a growth of the amplitude Δ and decreases nonlinearly with a growth of the period l. The waviness period has a stronger influence on the additional displacement than the amplitude. As it follows from calculations, the dependence $C(\overline{p})$ can be approximated by a power law only for small values of the nominal pressure $\overline{p}$.

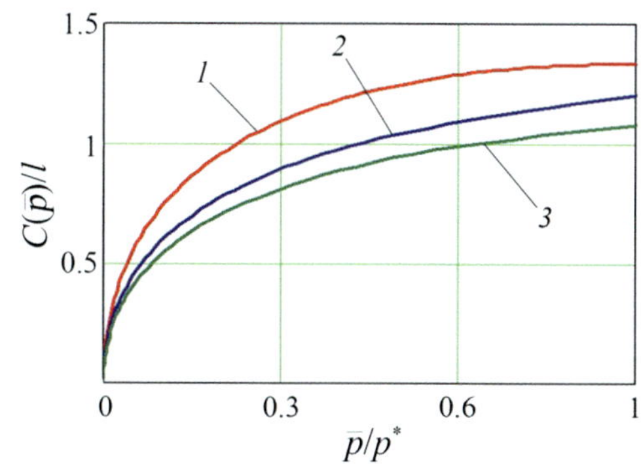

Figure 3.2 Additional displacement function divided by l for the wavy profile Eq. (3.10) at $m = 0$ (1); $m = 0.5$ (2); $m = 0.8$ (3).

The considered model is a one-scale model, however a real waviness (roughness) profile is multiscale. As a first approximation, it can be represented by a sum of sinusoids with various amplitudes and periods. Tsukanov (2018) considered a contact problem for a rigid body with a two-scale surface wavy profile and an elastic half-plane. The two-scale profile was modeled by the following function:

$$f(x) = \Delta_1\left(1 - \cos\frac{2\pi x}{l_1} + \frac{1}{k}\left(1 - \cos\frac{2\pi n x}{l_1}\right)\right). \tag{3.13}$$

Here, $k = \Delta_1/\Delta_2$ is the ratio of the amplitudes of the long-wave and short-wave harmonics of waviness; $n = l_1/l_2$ is the ratio of their periods. The graph of the waviness profile at $k = 10$ and $n = 11$ is presented in Fig. 3.3.

At $n/k < 0.5$, for calculating the normal pressure in periodic contact of the surface described by Eq. (3.13) and an elastic half-plane, the integral equation was obtained and solved by Tsukanov (2018) using a variable transform. The analytical expression for the contact pressure, obtained by Tsukanov (2018), has the following form:

$$p(x) = p_1(x) + p_2(x) \tag{3.14}$$

$$p_1(x) = \frac{\sqrt{2}\pi E \Delta_1}{(1-\nu^2) l_1}\left|\cos\frac{\pi x}{l_1}\right|\sqrt{\cos\frac{2\pi x}{l_1} - \cos\frac{2\pi a}{l_1}} \tag{3.15}$$

$$\begin{aligned} p_2(x) = &-\frac{2E}{(1-\nu^2)}\frac{\Delta_1 n}{l_1 k}\sqrt{1 - \left(\frac{\tan(\pi x/l_1)}{\tan(\pi a/l_1)}\right)^2} \\ &\times \sum_{j=1}^{\infty} U_{j-1}\left(\frac{\tan(\pi x/l_1)}{\tan(\pi a/l_1)}\right)\int_{-1}^{1}\frac{\phi(s,a)T_j(s)}{\sqrt{1-s^2}}ds \end{aligned} \tag{3.16}$$

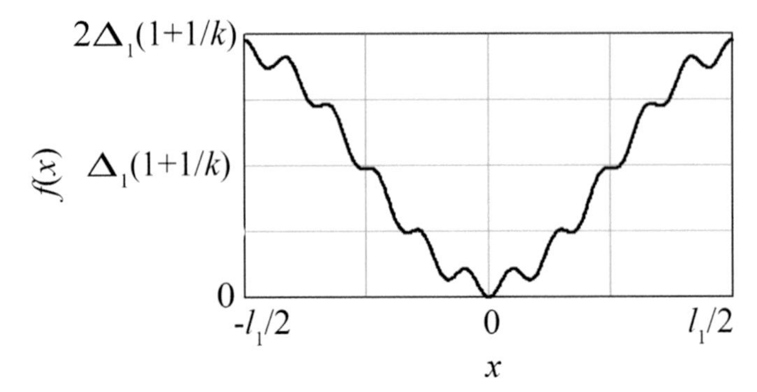

Figure 3.3 Example of the profile with two-scale waviness.

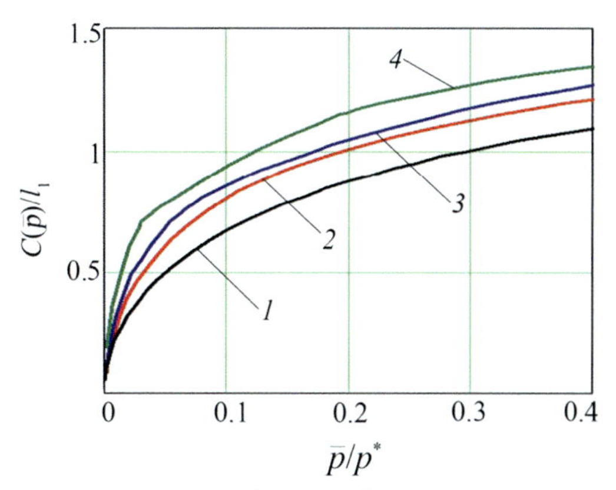

Figure 3.4 Additional displacement function for wavy profiles: single-scale profile (*curve 1*) and two-scale profiles Eq. (3.13) for $n = 5$ (*curve 2*), $n = 7$ (*curve 3*), and $n = 11$ (*curve 4*).

$$\phi(s, a) = \frac{\tan(\pi a/l_1)s}{1 + (\tan(\pi a/l_1)s)^2} U_{n-1}\left(\frac{1 - (\tan(\pi a/l_1)s)^2}{1 + (\tan(\pi a/l_1)s)^2}\right) \tag{3.17}$$

where U_n are Chebyshev polynomials of the second kind of n degree.

The additional displacement function Eq. (3.6) in the dimensionless form for a two-scale waviness at $\Delta_1 = 1$ mm, $l_1 = 10$ mm, and $k = 40$, which is calculated with the use of Eqs. (3.13)−(3.17) for various values of ratio n of the periods of short and long harmonics, is shown in Fig. 3.4 (Tsukanov, 2018).

It follows from the results (Fig. 3.4) that the main features of the additional displacement function for the two- and single-scale profiles are similar. Influence of the amplitude and period of the long-wave asperities has the same character as that for the single-scale waviness. However, existence of the short-wave harmonics leads to a discontinuity of the derivative of the additional displacement function $C(\overline{p})$. This effect increases with increasing of the ratio n of the periods of short and long harmonics. The ratio of the amplitudes k linearly influences the additional displacement, as it follows from Eqs. (3.15) and (3.16).

3.2 Additional displacement function (3-D analysis)

A similar approach is used to calculate the additional displacement function in contact between an elastic half-space ($z > 0$) and a rigid nominally flat body having a regular periodic relief in the directions of the x and y axes. The method of calculation of the contact characteristics for a periodic system of spherical punches penetrating into the elastic half-space that takes into account mutual influence of the contact spots is described in Chapter 1.

Let us consider the subregion Ω_0, which is not smaller than a period. Using the principle of localization (Chapter 1, Section 1.3.2) we can calculate the displacement $u_z(x, y)$ of the half-space surface loaded by the pressures $p_i(x, y)$ within contact spots ω_i taking into account the real pressures at the contact spots distributed in the vicinity of the point (x, y) in a region Ω_0 $(\omega_i \in \Omega_0)$ and replacing the real pressure distribution at remote contact spots $\omega_i \in \Omega \backslash \Omega_0$ with the nominal pressure $p(x, y)$, i.e.,

$$\frac{\pi E}{1-\nu^2} u_z(x, y) = \sum_{i=1}^{n} \iint_{\omega_i} \frac{p_i(\xi, \eta) d\xi d\eta}{\sqrt{(\xi - x)^2 + (\eta - y)^2}} - \iint_{\Omega \backslash \Omega_0} \frac{p(\xi, \eta) d\xi d\eta}{\sqrt{(\xi - x)^2 + (\eta - y)^2}}. \tag{3.18}$$

As it is shown in Chapter 1 for a periodic contact of spherical asperities penetrating into an elastic half-space, this substitution can be carried out with high degree of accuracy by choosing an appropriate region Ω_0, in which the real pressures at the contact spots are considered.

The elastic displacement $u_z(x, y)$ in the left-hand side of Eq. (3.18) satisfies the contact condition within the region Ω_0:

$$u_z(x, y) = D - f(x, y) + h(x, y). \tag{3.19}$$

Here D is the penetration, the function $f(x, y)$ describes the surface macroshape, and the function $h(x, y)$ describes the surface periodic microgeometry. From Eqs. (3.18) and (3.19) we obtain the following integral equation:

$$D - f(x, y) = C[p(x, y)] + \iint_{\Omega} \frac{p(\xi, \eta) d\xi d\eta}{\sqrt{(\xi - x)^2 + (\eta - y)^2}}. \tag{3.20}$$

Here the function $C[p(x,y)]$ takes the form

$$C[p(x,y)] = -h(x,y) + \frac{1-\nu^2}{\pi E}\left[\sum_{i=1}^{N}\iint_{\omega_i}\frac{p_i(\xi,\eta)\mathrm{d}\xi\mathrm{d}\eta}{\sqrt{(x-\xi)^2+(y-\eta)^2}} - \iint_{\Omega_0}\frac{p(x,y)\mathrm{d}\xi\mathrm{d}\eta}{\sqrt{(x-\xi)^2+(y-\eta)^2}}\right]. \quad (3.21)$$

This function describes the additional displacement due to microgeometry. As follows from Eq. (3.21), the additional displacement function depends on distribution of the contact pressure in real contact spots, which is calculated taking into account the shape and location of asperities in the subregion Ω_0 and also mutual influence of the contact spots.

Since the region Ω_0 is much smaller than the nominal contact area, in calculating the function $C[p(x,y)]$, it is possible to neglect the macro curvature of the contacting bodies, which is described by the function $f(x,y)$, and calculate the additional displacement at a particular point (x_0,y_0) of the nominal contact region by using the solution of periodic contact problems in which the asperities shape and space distributions model the surface microgeometry at the point (x_0,y_0). Under the assumption that the region Ω_0 in Eq. (3.21) is a circle of the radius R_0 and the pressure is distributed uniformly in this region and it is equal $p(x_0,y_0)$, Eq. (3.21) can be reduced to the form

$$C(p) = C[p(x_0,y_0)] = -h(x_0,y_0) + \frac{1-\nu^2}{\pi E}\left[\sum_{i=1}^{N}\iint_{\omega_i}\frac{p_i(\xi,\eta)\mathrm{d}\xi\mathrm{d}\eta}{\sqrt{(x_0-\xi)^2+(y_0-\eta)^2}} - 2\pi R_0 p(x_0,y_0)\right]. \quad (3.22)$$

As it follows from the periodic contact problem solutions (see Chapter 1), it is possible to derive the unique solution of this problem and to determine the additional displacement as a function of the surface microgeometry parameters and the nominal pressure.

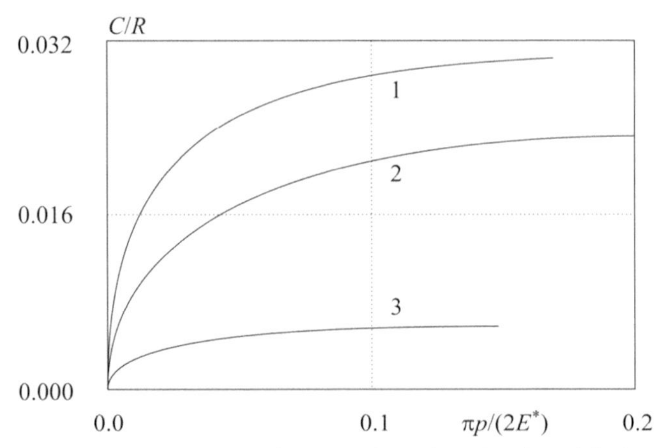

Figure 3.5 Additional displacement functions for a three-level model (1) and for one-level models with $l/R = 0.6$ (2) and $l/R = 0.3$ (3).

Fig. 3.5 illustrates the function $C(p)$ calculated from Eq. (3.22) for the periodic system of spherical indenters of radius R located at the sites of hexagonal lattice with pitch $l/R = 0.6$ and characterized by the following relative difference in heights of the asperities: $(h_0 - h_1)/R = 0.01$, $(h_0 - h_2)/R = 0.015$ (curve 1), and for the one-level system of spherical indenters located at the sites of hexagonal lattice with pitch $l/\ R = 0.6$ (curve 2) and $l/R = 0.3$ (curve 3). The functions $p_i(x, y)$ in Eq. (3.22) were calculated from Eq. (1.37) for the one-level model and from Eq. (1.46) for the three-level model of asperities height distributions. The results indicate that the rate of change of the function $C(p)$ decreases as the nominal pressure increases. If the real contact area is close to saturation, the additional displacement is close to the constant value, i.e., $dC/dp = 0$. This situation corresponds to the transition from the discrete to continuous contact. The results also indicate that the power function can be used to describe the additional displacement only for the low values of the nominal pressure.

3.3 Additional displacement due to microgeometry and adhesion

The function of additional displacement taking into account microgeometry and adhesion was constructed and analyzed by Goryacheva and Makhovskaya (2017a,b) based on the solution of the contact problem for two nominally flat surfaces, one of which was the surface of an elastic half-space, while the surface of the other was rigid and had a regular relief. The method for calculation of the contact characteristics for this problem was given in Chapter 2 by using the simplest version of the localization principle, in which the contact and adhesion of only one asperity with the half-space was taken into account, while the action of all other asperities was replaced by the nominal pressure distributed in the region $r \geq R_{\text{eff}}$, where $R_{\text{eff}} = l\left(\frac{\sqrt{3}}{2\pi}\right)^{1/2}$ for the hexagonal lattice with pitch l.

In the case of contact between the surfaces, the additional displacement is determined by Eq. (2.23) for $C = -d_a$. Assuming that all asperities have tops of spherical shape with the radius R, we set $n = 1$ and $A = 1/(2R)$ in Eq. (2.23) to obtain for the additional displacement:

$$C = \frac{a^2}{R} - \frac{2p_0 b}{E^*}\sqrt{1 - \frac{a^2}{b^2}} - \frac{2\overline{p}R_{\text{eff}}}{E^*}\sqrt{1 - \frac{a^2}{R_{\text{eff}}^2}}, \tag{3.23}$$

where p_0 is the adhesion pressure in accordance with the Maugis model of the molecular adhesion given by Eqs. (2.6) and (2.7), and $E^* = E/(1 - \nu^2)$. Eq. (3.23) does not completely define the dependence of the additional displacement C on the nominal pressure $\overline{p}$, since it contains two unknown values: the contact radius a and the adhesion radius b. So, we need two more equations. The first one follows from Eq. (2.24) taking into account Eq. (2.16), from which we have the equation defining the nominal pressure $\overline{p}$ through the values a and b:

$$\begin{aligned} \overline{p} = \frac{\pi}{\sqrt{3}l^2}\left[\frac{4E^* a^3}{3R} - 2p_0 b^2\left(\arccos\frac{a}{b} + \frac{a}{b}\sqrt{1 - \frac{a^2}{b^2}}\right)\right] \\ \left(\arccos\frac{a}{R_{\text{eff}}} + \frac{a}{R_{\text{eff}}}\sqrt{1 - \frac{a^2}{R_{\text{eff}}^2}}\right)^{-1} \end{aligned} \tag{3.24}$$

The second equation relating the contact radius a and the adhesion radius b follows from the adhesion condition given by Eq. (2.12) and relation for the gap given by Eq. (2.25):

$$\begin{aligned} h(b) = \left[\frac{b^2 - 2a^2}{2R} + \frac{2p_0 b}{E^*}\sqrt{1 - \frac{a^2}{R_{\text{eff}}^2}}\right]\frac{2}{\pi}\arccos\frac{a}{b} + \frac{a^2}{\pi R}\sqrt{\frac{b^2}{a^2} - 1} \\ - \frac{4p_0}{\pi E^*}(b - a) - \frac{4\overline{p}R_{\text{eff}}}{\pi E^*}\left[\mathbf{E}\left(\frac{b}{R_{\text{eff}}}\right) - \mathbf{E}\left(\frac{a}{b}, \frac{b}{R_{\text{eff}}}\right)\right] = \frac{w_a}{p_0}, \end{aligned} \tag{3.25}$$

where w_a is the specific work of adhesion. Eqs. (3.23)–(3.25) define the parametric dependence of the additional displacement C on the nominal pressure $\overline{p}$, where b is the parameter. At each value of b, Eq. (3.25) should be solved numerically for a.

In the case of no contact between the surfaces, where only adhesion attraction acts between them, the parametric dependence of the additional displacement C on the nominal pressure $\overline{p}$ follows from Eq. (2.27) with $n = 1$ and $A = 1/(2R)$:

$$C = \frac{b^2}{2R} - \frac{4b}{\pi E^*}\left[p_0 - \overline{p}\frac{b}{R_{\text{eff}}}\mathbf{E}\left(\frac{b}{R_{\text{eff}}}\right)\right] + \frac{w_a}{p_0},$$
$$\overline{p} = -\frac{2\pi p_0 b^2}{\sqrt{3}l^2}. \tag{3.26}$$

For calculations, we introduce the dimensionless additional displacement C^* and nominal pressure $\overline{p}^*$ by the relations:

$$C^* = \frac{4C}{3}\left(\frac{E^{*2}}{\pi^2 w_a^2 R}\right)^{1/3}, \quad \overline{p}^* = \frac{2\overline{p}}{3\pi w_a R \eta_s}, \tag{3.27}$$

where the scale parameter η_S defines the density of asperities, it equals $\eta_s = \frac{2}{\sqrt{3}l^2}$ for the case of the hexagonal lattice. Besides, we introduce the dimensionless radii of contact and adhesion:

$$\alpha = a\left(\frac{E^*}{\pi w_a R^2}\right)^{1/3}, \quad \beta = b\left(\frac{E^*}{\pi w_a R^2}\right)^{1/3} \tag{3.28}$$

and the following two dimensionless parameters

$$\lambda = p_0\left(\frac{9R}{2\pi w_a E^{*2}}\right)^{1/3}, \tag{3.29}$$

$$L_1 = l\left(\frac{\sqrt{3}}{2\pi}\right)^{1/2}\left(\frac{E^*}{\pi w_a R^2}\right)^{1/3}. \tag{3.30}$$

The adhesion parameter λ given by Eq. (3.29) was suggested for the description of spherical adhesive contact (Maugis, 1992). For the Lennard-Jones law of molecular attraction given by Eq. (2.3), the maximum adhesion pressure is calculated as follows:

$$p_m = \frac{16}{9\sqrt{3}}\frac{w_a}{z_0}. \tag{3.31}$$

By setting $p_0 = p_m$ in the Maugis model of adhesion defined by Eqs. (2.6) and (2.7), one can establish the relation between the adhesion parameter λ and the Tabor parameter μ_T defined by Eq. (2.5):

$$\lambda = \frac{16}{9^{2/3}3^{1/2}(2\pi)^{1/3}}\mu_T \approx 1.1\mu_T. \tag{3.32}$$

The parameter L_1 in Eq. (3.30) characterizes the mutual influence of asperities. At large L_1, the mutual influence is insignificant. The solution obtained is valid for $L_1 > \beta$.

For the calculations, we use Eqs. (3.24)–(3.26), which allow us for the given values of the parameters λ and L_1 to calculate the dimensionless real contact area $A_1 = \pi\alpha^2$ and dimensionless additional displacement C^* as functions of the dimensionless nominal pressure $\overline{p}^*$. The results obtained are presented in Fig. 3.6. Solid lines correspond to a relatively dense contact ($L_1 = 5$), and dashed lines correspond to the case where mutual influence of asperities is insignificant ($L_1 = 50$). The curves presented were calculated for three values of the parameter λ: $\lambda = 0.1$ (curves 1), $\lambda = 0.5$ (curves 2), and $\lambda = 2$ (curves 3).

In Fig. 3.6a, graphs of the dimensionless real contact area as a function of the dimensionless nominal pressure are presented. In Fig. 3.6b, graphs of the dimensionless additional displacement as a function of the dimensionless nominal pressure are given.

The results indicate that due to adhesion, the surfaces are in contact not only for positive (compressive), but also for negative (tensile) nominal pressures, and the absolute values of these negative pressures are higher for larger values of the adhesion parameter λ. Analysis of the calculated results shows that the adhesion parameter λ significantly influences the additional displacement function $C^*\left(\overline{p}^*\right)$. The parameter L_1 characterizing the mutual influence of asperities has only slight effect on this function for small λ. As λ increases, this effect becomes more significant. The quantity $C^*(0) = C_0^*$, at which the nominal contact pressure $\overline{p}^*$ changes its sign, defines the equilibrium displacement.

When the surfaces are in contact under a positive pressure and then the pressure is starting to decrease up to negative values, the contact breaks apart at some value of the negative pressure. This value (pull-off pressure) $\overline{p}^*_{\min}$ corresponds to the far-left point of the graphs in Fig. 3.6b.

In Fig. 3.7, the absolute value of the dimensionless pull-off pressure is presented as a function of the adhesion parameter. The solid line

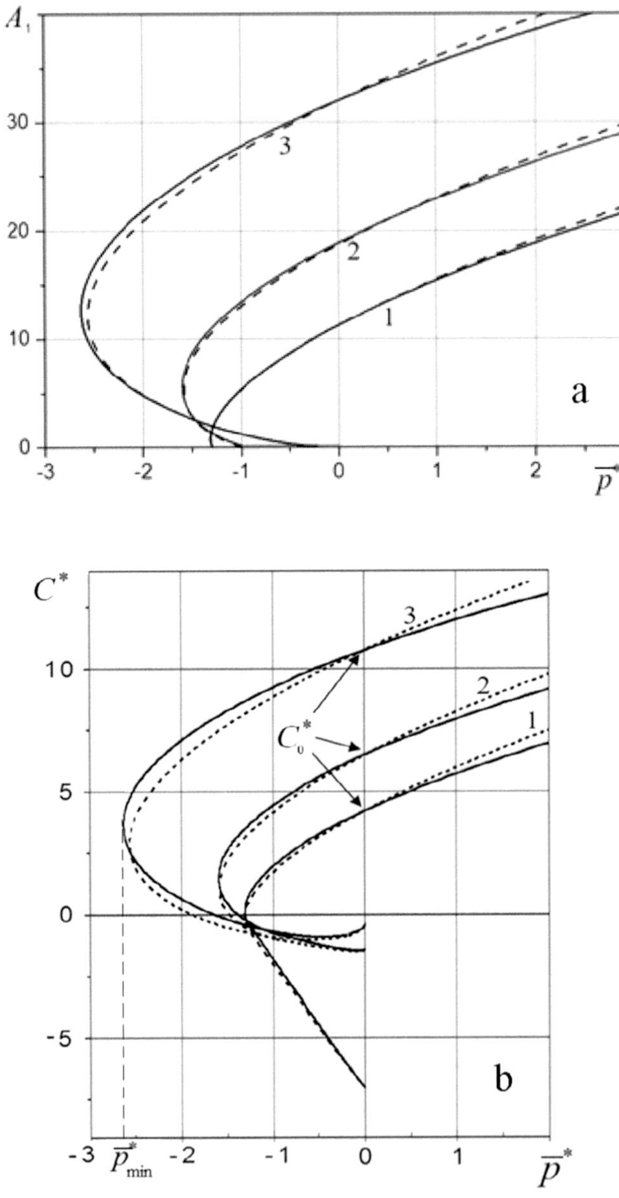

Figure 3.6 Real contact area (a) and additional displacement (b) versus nominal pressure for $\lambda = 0.1$ (*curves 1*), $\lambda = 0.5$ (*curves 2*), and $\lambda = 2$ (*curves 3*) and for $L_1 = 5$ (*solid lines*) and $L_1 = 50$ (*dashed lines*).

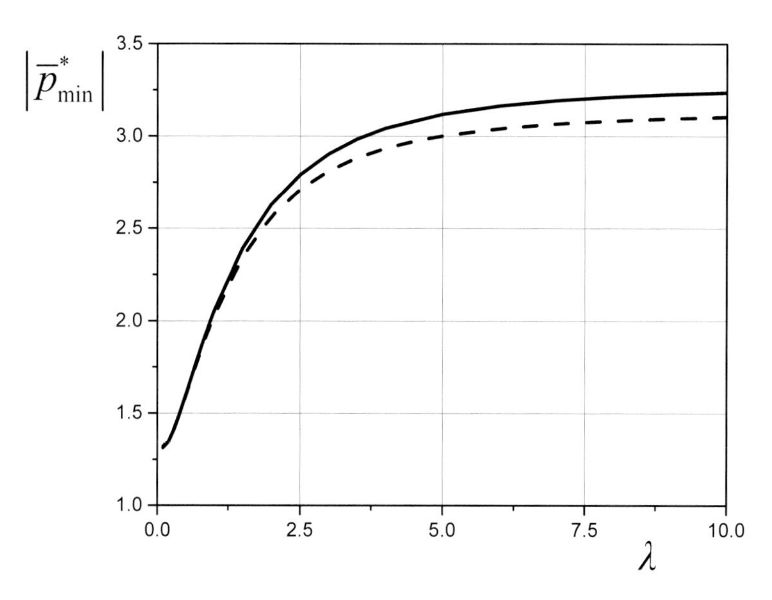

Figure 3.7 Pull-off pressure versus the adhesion parameter λ for $L_1 = 5$ (*solid line*) and $L_1 = 50$ (*dashed line*).

corresponds to $L_1 = 5$ and the dashed one to $L_1 = 50$. It follows from the results that an increase in the contact density leads to an increase in the pull-off pressure, which is particularly significant for large λ. Thus, surfaces with denser arranged asperities are able to stay in contact under higher tensile pressures, the adhesion and elastic properties being the same.

3.4 Effective work of adhesion for surfaces with microgeometry

Similar to the specific work of adhesion for smooth flat surfaces w_a defined by Eq. (2.4), one can calculate the specific work of adhesion w_{rough} for nominally flat rough surfaces, which is necessary to retract the surfaces from the equilibrium distance $d^* = d_0^*$ to infinity (the distance between the surfaces is defined as negative additional displacement $d = -C$):

$$w_{\text{rough}} = -\frac{9}{8}\pi w_a R\eta_s \left(\frac{\pi^2 w_a^2 R}{E^{*2}}\right)^{1/3} \int_{d_0^*}^{\infty} p^*(z)dz$$

$$= \frac{9}{8}\pi w_a R\eta_s \left(\frac{\pi^2 w_a^2 R}{E^{*2}}\right)^{1/3} w^*_{\text{rough}}. \quad (3.33)$$

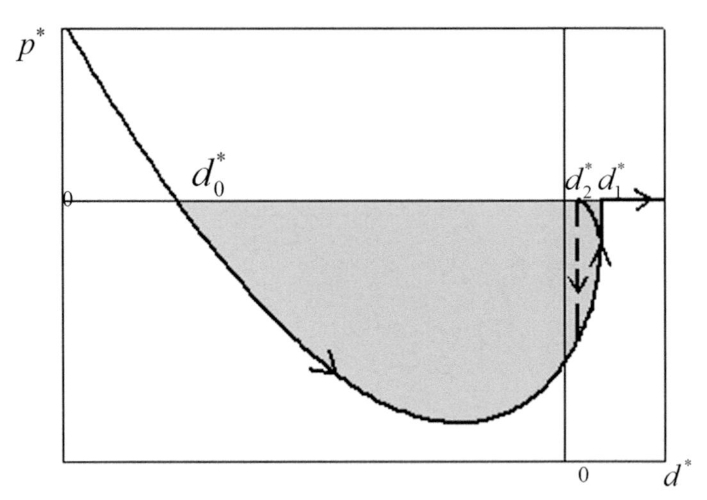

Figure 3.8 Dimensionless nominal pressure as a function of distance between two nominally flat surfaces with microgeometry and adhesion.

The quantity w^*_{rough} can be considered as dimensionless specific work of adhesion. Numerically it equals the dashed area between the plot of the function $p^* = p^*(d^*)$ for negative p^* and the axis $p^* = 0$ (Fig. 3.8).

Note that the work that must be done to separate the surfaces is generally greater than the work done by the adhesive forces when the same surfaces approach from infinity to the equilibrium distance; i.e., a hysteresis takes place. This follows from the ambiguity of the curves of the nominal pressure versus the distance, which takes place at sufficiently large values of the adhesion parameter $\boldsymbol{\lambda}$. When the surfaces are separated from each other, the contact break occurs at $d^* = d_1^*$, and when they approach, they snap into contact at $d^* = d_2^*$ (Fig. 3.8). Thus, the values of the effective specific work of adhesion of two rough surfaces in the presence of adhesion is different for approach and retraction. The difference between the values of the work of adhesion in approach and retraction of smooth elastic bodies in an axisymmetric contact, which equals the energy dissipation during cyclic approach-retraction of the bodies, is calculated and investigated in Chapter 4. Relation Eq. (3.33), which is used for the calculation below, determines the effective specific work of adhesion in retraction of the surfaces from each other.

In Fig. 3.9, the dependencies of the effective work of adhesion w^*_{rough} on the adhesion parameter λ are presented. As previously, the solid line corresponds to $L_1 = 5$ and the dashed one to $L_1 = 50$. The results show that as

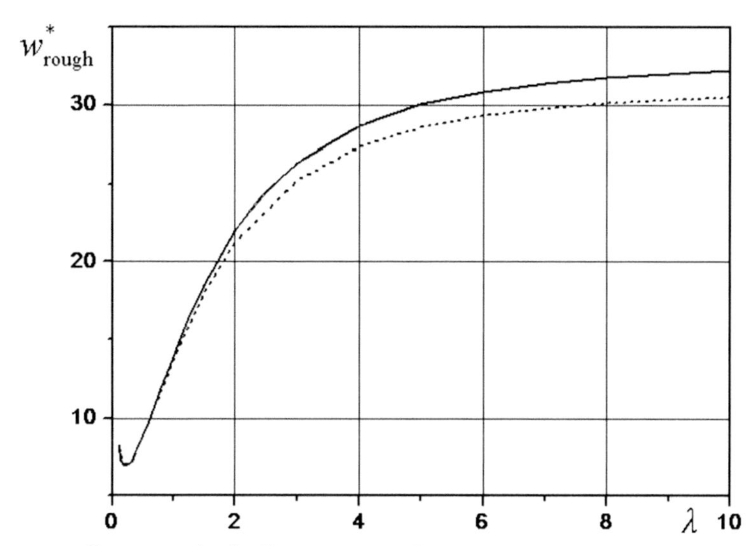

Figure 3.9 Effective work of adhesion versus the adhesion parameter for $L_1 = 5$ (*solid line*) and $L_1 = 50$ (*dashed line*).

the density of asperities on a surface increases, the effective work of adhesion of such surface increases, the increase being more significant for higher λ. As the adhesion parameter λ increases, the effective work of adhesion w^*_{rough} first slightly decreases and then increases. Note that as $\lambda \to 0$, the dimensionless width of the adhesion region $\beta \to \infty$, so for very small λ the condition $L_1 > \beta$, under which the model is applicable, is not satisfied: separate regions of adhesion merge into one region.

In the general case, for arbitrary values of the parameters λ and L_1, the effective work of adhesion w^*_{rough} is obtained only by numerical integration. For $\lambda \to \infty$ from Eqs. (3.23)–(3.25) taking into account that $d = -C$, we obtain the parametric relation between the dimensionless nominal pressure and distance between the surfaces:

$$p^* = \frac{\dfrac{4\alpha}{9} - \dfrac{2\sqrt{2}\pi\alpha^{\frac{3}{2}}}{3}}{\arccos\dfrac{\alpha}{L_1} + \dfrac{\alpha}{L_1}\sqrt{1 - \dfrac{\alpha^2}{L_1^2}}}$$

$$d^* = \frac{4(\alpha^2 + \sqrt{2\alpha})}{3} + \frac{8}{9\pi L_1}\frac{2\alpha^3 - 3\sqrt{2}\pi\alpha^{3/2}}{\arccos\dfrac{\alpha}{L_1} + \dfrac{\alpha}{L_1}\sqrt{1 - \dfrac{\alpha^2}{L_1^2}}}\sqrt{1 - \frac{\alpha^2}{L_1^2}} \tag{3.34}$$

The dependencies given by Eq. (3.34) still do not allow to take the integral in Eq. (3.33) analytically, but they considerably simplify the calculation. By substituting Eq. (3.34) into Eq. (3.33) and calculating the integral numerically, we obtain the effective work of adhesion w^*_{rough} as a function of the parameter L_1, which characterizes the density and mutual influence of asperities, for the case $\lambda \to \infty$. This function is plotted in Fig. 3.10.

The limit value of the effective work of adhesion w^*_{rough} can be obtained in the closed form for $L_1 \to \infty$, i.e., disregarding the mutual influence of asperities. By setting $L_1 \to \infty$ in Eq. (3.34), one obtains the relations of the Maugis model for the single spherical contact (Maugis, 1992):

$$p^* = \frac{8\alpha^3}{9\pi} - \frac{4\sqrt{2}\alpha^{3/2}}{3}, \quad d^* = -\frac{4\alpha^2}{3} + \frac{4\sqrt{2}\alpha}{3}. \tag{3.35}$$

In this case, the effective specific work of adhesion can be calculated by taking the integral:

$$w^*_{\text{rough}} = \int_{\alpha_1}^{\alpha_2} p^*(\alpha) d_0'(\alpha) d\alpha, \tag{3.36}$$
$$\text{where } \alpha_1 = 1/2, \quad \alpha_2 = (6\pi)^{2/3}/2.$$

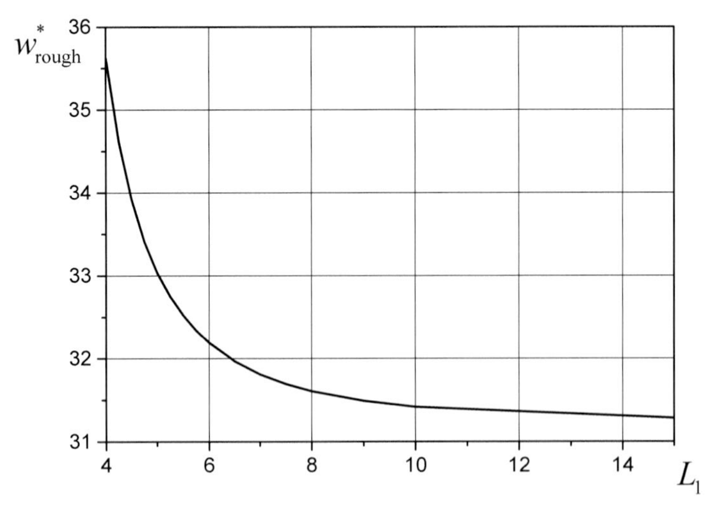

Figure 3.10 Effective work of adhesion as a function of the asperities density parameter at $\lambda \to \infty$.

From Eq. (3.36) we finally have

$$w^*_{\text{rough}} = \frac{2}{315\pi}\left(15\pi - 1 + 216\pi^{10/3}6^{1/3} - 90\pi^{7/3}6^{1/3}\right) \approx 31,34. \quad (3.37)$$

Note that the adhesion parameter λ, as previously shown by Maugis (1992), in two limit cases ($\lambda \to 0$ and $\lambda \to \infty$) leads to the classical simplified models of adhesion for two smooth axisymmetric bodies. The case $\lambda \to 0$ corresponds to the DMT model (Derjaguin et al., 1975), which is applicable for sufficiently hard interacting bodies. This model takes into account the tensile stresses caused by adhesive forces outside the contact area, but it is considered that they do not affect the stress distribution within the contact area. Another classical model—the JKR model (Johnson et al., 1971)—corresponds to the limit case $\lambda \to \infty$, and it is applicable for sufficiently soft bodies and relatively high values of their surface energy. In this model, the solution of the contact problem is obtained taking into account tensile stresses inside the contact area near its edges and zero outside the contact area, while the maximum compressive stresses in the contact are higher than they would be without taking into account adhesion.

The obtained solution Eqs. (3.23)–(3.25) with the parametrization Eqs. (3.27)–(3.30) tends to the JKR solution as $\lambda \to 0$ and $L_1 \to \infty$, and it tends to the DMT solution as $\lambda \to 0$ and $L_1 \to \infty$. For the case $\lambda \to \infty$ and arbitrary L_1, relations Eq. (3.34) are a generalization of the JKR solution to the case of multiple contact.

It is important to note that the effective specific work of adhesion for nominally flat rough surfaces, defined by Eq. (3.33), allows an approximate estimate of the contact characteristics for the adhesive contact of rough bodies by using the relations for smooth bodies. This definition of the specific work of adhesion w_{rough} makes it possible to use existing models of adhesive contact of smooth elastic bodies to calculate the nominal contact characteristics. It is sufficient to replace the specific work of adhesion w_a for smooth bodies by the value w_{rough} in these models. The disadvantage of such estimates is that the deformations of the rough layer itself are not taken into account at the macrolevel; they are taken into account only when calculating w_{rough}. These deformations significantly affect the distribution of nominal contact stresses, the size of the nominal contact areas, and the nominal gap between bodies after deformation, even in the absence of adhesion. Therefore, for a more accurate calculation of the characteristics of the adhesive contact of rough bodies with a given macroscopic shape, it is necessary to use the method of constructing integral equations at the macrolevel, taking into account the function of additional displacement due to deformation of roughness. In this case, the compliance of the surface rough layer should be described by relations Eqs. (3.23)–(3.26).

3.5 Two-scale analysis of the contact problems

In contact of deformable bodies with rough surfaces, there are at least two length scales (Fig. 3.11): one scale relates to the characteristic contact length L, and the other one relates to the characteristic size and distance between asperities, which can be characterized by the average scale parameter l. The ratio l/L is variable in contact interaction.

Under low contact pressures, it is conceivable that $L \sim l$; i.e., there are a finite number of asperities in the contact. In this case, the method described in Chapter 1 (see Section 1.4) can be used to determine the contact characteristics (the nominal and real contact area, the load distribution between contact spots, the real pressure distribution, etc.).

If $l/L \ll 1$, the shape of the surface macrogeometry $f(x, y)$ within the region $\Omega_0 << \Omega$ can be neglected. It can be assumed that a uniformly distributed pressure $p(x, y)$ is applied in the subregion Ω_0. Then, as it was proved in Section 3.2 by using the principle of localization, the additional displacement due to the rough layer can be calculated from Eq. (3.22).

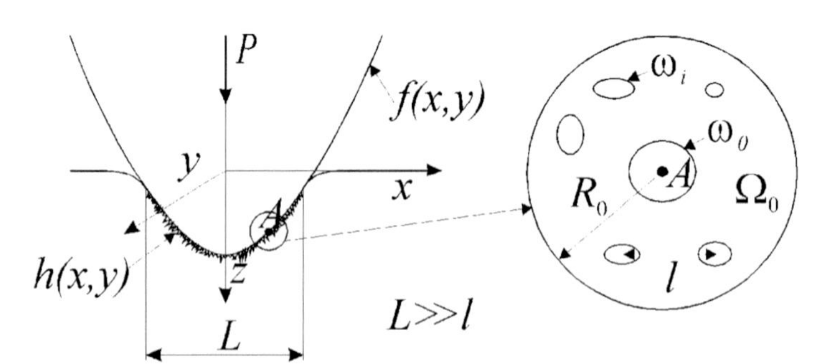

Figure 3.11 Scheme of contact of bodies with given macroshape $f(x, y)$ and microgeometry $h(x, y)$.

3.5.1 Formulation of the contact problems at macroscale using the additional displacement function

The additional displacement function $C(p)$ calculated for measured microgeometry parameters (average curvature of asperities and contact density, as well as asperities height distribution) and characterizing the compliance of the rough surface layer is used to formulate the contact problem at macroscale. The method of its calculation is described in Sections 3.2–3.4.

As shown in Section 3.2, the integral Eq. (3.20) can be used to calculate the contact pressure $p(x, y)$ at macroscale taking into account the additional displacement $C[p(x, y)]$ due to surface microgeometry. If two elastic bodies come into contact, this equation takes the form:

$$C[p(x,y)] + \frac{1}{\pi E^*}\iint_{\Omega} \frac{p(x,y)dx'dy'}{\sqrt{(x-x')^2+(y-y')^2}} = D - f_1(x,y) - f_2(x,y), \quad (x,y) \in \Omega. \tag{3.38}$$

Here, Ω is the nominal contact area, $f_i(x, y)$ $(i = 1, 2)$ are the surface macroshapes of the contacting bodies, D is their approach, and $E^* = \left(\frac{1-\nu_1^2}{E_1} + \frac{1-\nu_2^2}{E_2}\right)^{-1}$, where E_i and ν_i are the Young modulus and Poisson ratio characterizing the elastic properties of the contacting bodies. If the functions $f_i(x, y)$ are smooth, the additional condition at the boundary $\partial\Omega$ of the contact area must be added:

$$p(x,y) = 0, \quad \text{if } (x,y) \in \partial\Omega. \tag{3.39}$$

If the load P applied to the contact area is known, the equilibrium equation is used to calculate the unknown approach D:

$$P = \iint_{\Omega} p(\xi,\eta)d\xi d\eta. \tag{3.40}$$

The methods to solve Eq. (3.38) with the additional conditions at the contact boundary were developed (Alexandrov and Kudish, 1979; Galanov, 1985; Goryacheva, 1979; Popov and Savchuk, 1971; Shtaierman, 1949) for different types of the function $C[p(x, y)]$ and different kernels $K(x, y, x', y')$ of the integral operator that are typical for the contact problems. In what follows, we give the solutions of some particular contact problems.

3.5.2 Analysis of the microgeometry effect on the nominal contact characteristics

As it was indicated in the previous sections, the additional displacement due to surface roughness (see Sections 3.1–3.3) and adhesion (see Section 3.4) can be calculated analytically for certain types of surfaces with regular microrelief. It was also concluded from the numerical calculations that the dependence of the additional displacement due to roughness on the nominal pressure under the condition of small displacements can be described by a power function.

Let us consider the contact of a strip punch or a long elastic cylinder, with an elastic layer of the thickness h ($|x| < \infty, \; 0 < z < h$), lying on a rigid foundation (Fig. 3.12).

This problem can be studied in 2-D formulation. The indenter macroshape is given by the function $z = f(x)$. The load P is applied to the indenter in the z-axis direction. The shear stresses within the contact region are negligibly small. We analyze two types of contact conditions at the layer/foundation interface ($z = h$):

there is no friction at the layer/foundation interface, i.e.,

$$\tau_{xz}(x, h) = 0, \quad u_z(x, h) = 0, \quad |x| < +\infty, \tag{3.41}$$

or the layer is bonded with the foundation, i.e.,

$$u_x(x, h) = u_z(x, h) = 0. \tag{3.42}$$

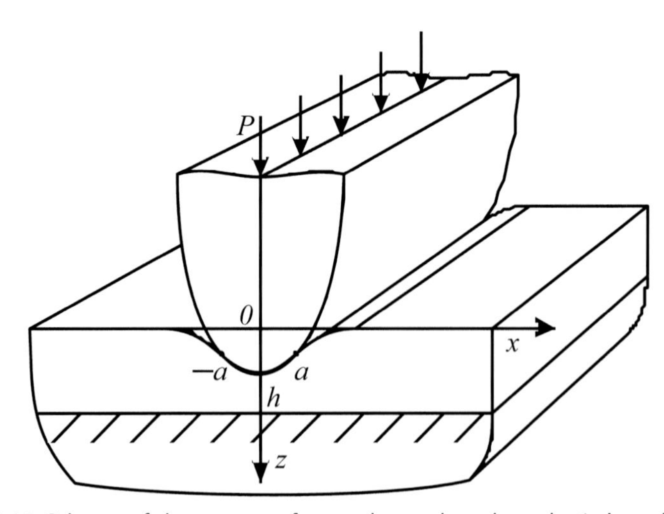

Figure 3.12 Scheme of the contact of a rough punch and an elastic layer lying on a rigid foundation.

The boundary conditions at the surface $z = 0$ are

$$\begin{aligned} \tau_{xz}(x,0) &= 0, \ \ u_z(x,0) = D - f(x), \quad |x| < a \\ \tau_{xz}(x,0) &= 0, \ \ \sigma_z(x,0) = 0, \quad a < |x| < \infty. \end{aligned} \tag{3.43}$$

The integral Eq. (3.38) takes the following form:

$$C[p(x)] + \frac{2(1-\nu^2)}{\pi E} \int_{-a}^{a} k\left(\frac{\xi - x}{h}\right) p(\xi)d\xi = D - f(x). \tag{3.44}$$

The kernel $k(t)$ of the integral operator in Eq. (3.44) has the form (Vorovich et al., 1974):

$$k(t) = \int_0^{+\infty} \frac{L(u)}{u} \cos utdu. \tag{3.45}$$

The form of the function $L(u)$ depends on the boundary conditions at $z = h$. If there is no friction at the interface $x = h$, then

$$L(u) = \frac{\cosh 2u - 1}{\sinh 2u + 2u}. \tag{3.46}$$

If the strip is bonded to the foundation at $x = h$, then

$$L(u) = \frac{2(3-4\nu)\sinh 2u - 4u}{2(3-4\nu)\cosh 2u + 4u^2 + 1 + (3-4\nu)^2}. \tag{3.47}$$

In dimensionless form Eq. (3.44) and the equilibrium condition Eq. (3.40) can be rewritten as

$$\int_{-1}^{1} k(\lambda(t - x_1))p_1(t)dt + C_1(p_1(x_1)) = \delta - f_1(x_1), \tag{3.48}$$

$$\int_{-1}^{1} p_1(x)dx = P_1. \tag{3.49}$$

Here, the following notations have been introduced:

$$\begin{aligned} &x_1 = \frac{x}{a}, \ \delta = \frac{D}{a}, \ \lambda = \frac{a}{h}, \ f_1(x_1) = \frac{f(ax_1)}{a}, \ P_1 = \frac{2(1-\nu^2)P}{\pi E a}, \\ &p_1(x_1) = \frac{2(1-\nu^2)}{\pi E} p(ax_1), \ \ C_1(p_1(x_1)) = \frac{C(p(ax_1))}{a}. \end{aligned} \tag{3.50}$$

If the contact half-width a is known in advance, Eqs. (3.48) and (3.49) form the complete system of equations for the determination of the dimensionless contact pressure $p_1(x_1)$ and the penetration δ. Eq. (3.48) is an equation of Hammerstein type. It can be reduced to the canonical form by introducing the following function:

$$\psi(x_1) = C_1(p_1(x_1)) + f_1(x_1) - \delta. \tag{3.51}$$

Then we have

$$\int_{-1}^{1} k(\lambda(t - x_1))C_1^{-1}(\psi(t) - f_1(t) + \delta)dt + \psi(x_1) = 0. \tag{3.52}$$

Here, $C_1^{-1}(x)$ is the inverse function to $C_1(x)$.

Eq. (3.52) can be solved by iteration. The function $\psi_0(x_1) = 0$ is used as an initial approximation of the solution. Subsequent approximations are calculated from the following recurrence relation:

$$\psi_{n+1}(x_1) = -\int_{-1}^{1} k(\lambda(t - x_1))C_1^{-1}(\psi_n(t) - f_1(t) + \delta)dt. \tag{3.53}$$

Convergence of the method depends on the type on the function $C_1(x)$. In Section 3.2 it was shown that for low nominal pressure, when the real contact area is far away from saturation, this function can be approximated by a power function, i.e.,

$$C[p(x, y)] = Bp^{\kappa}(x, y). \tag{3.54}$$

In this case, as it was proved by Goryacheva (1979), the method converges to the unique solution of Eq. (3.52) provided that the problem parameters satisfy the following inequality:

$$\left(\frac{1}{\kappa B_1^{1/\kappa}}\right)^2 \int_{-1}^{1} \left[\delta - \tilde{f}(x)\right]^{2/\kappa - 2} \times \left[\int_{-1}^{1} k^2(\lambda(t - x))dt\right] dx < 1. \tag{3.55}$$

Here,

$$B_1 = \frac{B}{2a}\left[\frac{\pi E}{2(1 - \nu^2)}\right]^{\kappa}. \tag{3.56}$$

For other ranges of the parameters, Eq. (3.48) can be solved by the Newton-Kantorovich method. Then for a given penetration D, the dimensionless contact pressure is calculated from Eq. (3.51). If D is unknown, the equilibrium equation is used to determine the penetration D for a known contact load P.

It is worth noting that if $C(p)$ is described by the power function Eq. (3.54), the pressure does not tend to infinity at the ends of the nominal contact region. To prove it, let us assume that the pressure has an integrable singularity of the type $(1-\xi)^{-\theta}$ $(0<\theta<1)$ at $\xi=1$. Then it follows from Eq. (3.48) that the left-hand side of this equation has the singularity $(1-\xi)^{-\kappa\theta}$, whereas there is no singularity at the right-hand side of this equation. This contradiction proves the proposition mentioned above. Thus, the additional displacement due to existence of surface microgeometry leads to the disappearance of the contact pressure singularities for the bodies with macroshapes that provide a discontinuity of the derivative of the surface displacement $u_z'(x)$ at the ends of the contact region.

For smooth macroshapes of the contacting bodies, there is no discontinuity of the function $u_z'(x)$ at the ends of the contact region, so the dimensionless contact pressure satisfies the following condition: $p_1(-1)=p_1(1)=0$. Taking into account this condition and also the condition $C(0)=0$, Eq. (3.48) can be reduced to the form:

$$\int_{-1}^{1}(k(\lambda(t-x_1))-k(\lambda(t-1)))p_1(t)dt+C_1(p_1(x_1))=f_1(1)-f_1(x_1). \tag{3.57}$$

Let us prove that the contact pressure $p_1(x_1)$ has zero derivatives at the ends of the contact region, i.e., $p_1'(-1)=p_1'(1)=0$, if the function $C(p)=Bp^{\kappa}$, $0<\kappa<1$. Upon differentiating Eq. (3.57) with respect to x_1 and setting $x_1=-1$ (the case $x_1=1$ can be analyzed in a similar manner), we obtain

$$\int_{-1}^{1}(-\lambda k'(\lambda(t-x_1)))p_1(t)dt+B_1\kappa p_1^{\kappa-1}(-1)p_1'(-1)=-f_1'(-1), \tag{3.58}$$

where B_1 is determined by Eq. (3.56). Since the function $p_1(x_1)$ is continuously differentiable, $p_1(-1)=p_1(1)=0$, and the kernel $k(t)$ of Eq. (3.45) is presented as $k(t)=-\ln|t|+F(t)$ (Goryacheva, 1979), where $F(t)$ is an analytical function, the integral term on the left-hand side of Eq. (3.58) is bounded. Since the value $f_1'(-1)$ on the right-hand side of Eq. (3.58) is also bounded, the second term in the left-hand side of this equation has to be bounded. For $0<\kappa<1$, it is possible only if $p_1'(-1)=0$.

As an example, let us consider the plane frictionless contact problem for a punch with flat base and a thick elastic layer with rough surface. So we use Eq. (3.48) for calculation of the contact pressure distribution taking into account $f(\xi) = 0$. The kernel of the integral Eq. (3.48) for the thick layer $(a/h \leq 1/2)$ can be approximated by the function $k(t) = -\ln|t| + a_0$ where $a_0 = -0.352$ if layer lies on the rigid foundation without friction, and $a_0 = -0.527$ for $\nu = 0.3$ if layer is bonded with the foundation (Vorovich et al., 1974).

Solving Eq. (3.52) by iteration, we obtain the following expression for the nominal contact pressure:

$$p_1(\xi) = B_1^{-1/\kappa}[\psi(\xi) + \delta]^{1/\kappa}, \tag{3.59}$$

where $\psi(\xi)$ is the limit of the function sequence $\{\psi_n(\xi)\}$ determined by

$$\begin{aligned} \psi_{n+1}(\xi) &= B_1^{-1/\kappa} \int_0^1 (\ln|t - \xi| + c_0)[\psi_n(t) + \delta]^{1/\kappa} dt, \\ c_0 &= \ln\left(\frac{a}{h}\right) - a_0. \end{aligned} \tag{3.60}$$

This limit exists if the condition Eq. (3.55) holds, which has the following form for this particular case:

$$\frac{\delta^{\frac{2}{\kappa}-2}\left(2c_0^2 - 3c_0 + 3.5\right)}{\kappa^2 B_1^{\frac{2}{\kappa}}} < 1. \tag{3.61}$$

Fig. 3.13 illustrates the pressure distributions in contact of a punch with flat base and a thick elastic layer for various values of parameter B characterizing the additional displacement Eq. (3.54) due to the punch surface roughness. In calculations we used $\kappa = 0.4$, $c_0 = 0.4$, and the dimensionless load $P_1 = 0.41 \cdot 10^{-2}$. The curves 1 and 2 were calculated for various values of parameter B, and the dashed curve corresponds to the pressure distribution for a smooth punch $(B = 0)$ with the same macroshape.

The results indicate that the contact pressure increases at the ends of the contact region if the parameter B decreases (i.e., if the roughness of the punch decreases), and it tends to infinity at the ends of the contact region for a smooth surface of the punch (*dashed line* in Fig. 3.13). Under the same load, the penetration of the punch depends on its roughness: for $B_1 = 0.75$

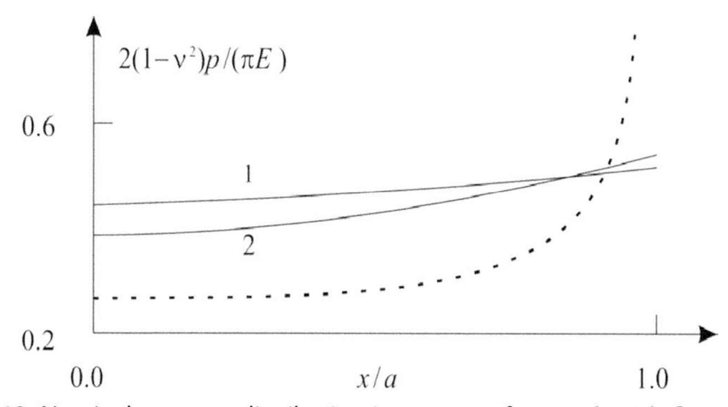

Figure 3.13 Nominal pressure distribution in contact of a punch with flat rough base and a thick elastic strip at $P_1 = 0.41 \cdot 10^{-2}$ for various roughness parameters: $B_1 = 0.75$ (*curve 1*) and $B_1 = 0.35$ (*curve 2*); *dashed line* corresponds to a punch with smooth surface.

the punch penetration is $\delta = 0.1$, and for a smoother surface $(B_1 = 0.35)$ the penetration is smaller $(\delta = 0.06)$.

In contact of a cylinder whose macroshape is described by the function $f(x) = x^2/(2R_0)$ (R_0 is the cylinder radius) and a thick elastic layer (see Fig. 3.12), the nominal contact pressure is zero at the ends of the contact area.

Fig. 3.14 illustrates the nominal pressure distribution for three various models of surface microgeometries considered in Chapter 1 (one-level models with various contact densities and a three-level model). The additional displacement functions for these models are presented in Fig. 3.5. The nominal pressure distributions were calculated for the given value of the dimensionless load $\overline{P} = 2(1-\nu^2)P/(\pi E R_0)$, applied to cylinder. The calculated contact half-widths for the considered microgeometry models are $a/R_0 = 0.09$ (1), $a/R_0 = 0.08$ (2), $a/R_0 = 0.065$ (3).

A similar method of analysis was used by Goryacheva (1979) to study the axisymmetric contact problem for an axisymmetric indenter with rough surface penetrating into an elastic half-space. The results indicate that due to roughness of the indenter, the contact pressures are more uniformly distributed within the contact region.

3.5.3 Combined effect of microgeometry and adhesion on the nominal characteristics of a two-scale contact

Eq. (3.20) was solved numerically for the case of contact between a smooth rigid punch and a rough elastic half-space, the roughness of which

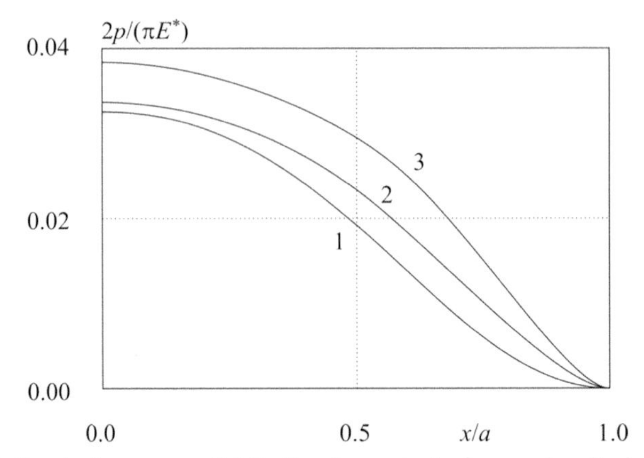

Figure 3.14 Nominal pressure distribution in contact of a rough cylinder and a thick elastic layer for various microgeometry parameters.

is replaced by a surface layer with properties specified by Eqs. (3.23)–(3.26). The calculations were carried out for various shapes of interacting bodies at macrolevel and various parameters of roughness L_1 and adhesion λ at microlevel. The shape of the punch at macrolevel is considered to be axisymmetric and is described by the power function $f(x) = A|x|^s$, $s \geq 1$.

Fig. 3.15 shows the distributions of nominal contact pressures for the punches of various shapes at macrolevel: conical $s = 1$, parabolic

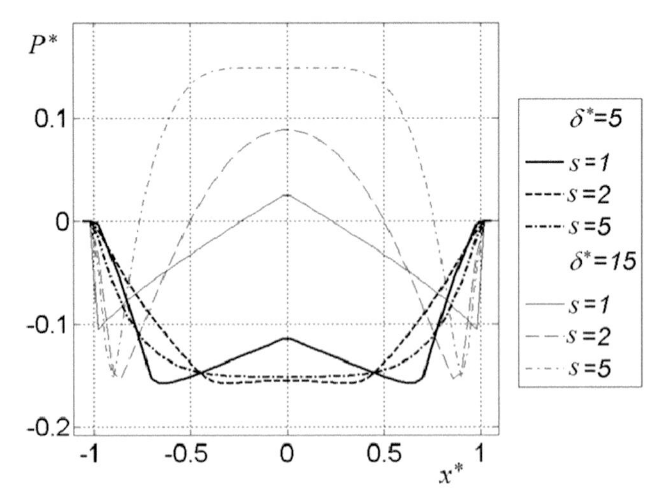

Figure 3.15 Distribution of dimensionless nominal contact pressure for two different values of the approach and for punches of various shapes at macrolevel.

characterized by s = 2 and $s = 5$, and at various values of the dimensionless approach of the bodies $\delta^* = 5$ and $\delta^* = 15$. At the microlevel, the following values are chosen for the parameters of adhesion and roughness: $\lambda = 0.1$, $L_1 = 5$. The results indicate that taking into account adhesion and roughness at microlevel leads to significant changes in the distribution of the nominal pressure: areas of negative pressure appear near the boundary of the contact region, and these areas are wider for smaller approach of the bodies. The distribution of contact pressure is significantly influenced not only by the magnitude of the approach of the contacting bodies, but also by the shape of the punch.

The results obtained substantially depend also on the adhesion parameter λ. Taking into account the mutual influence of asperities (parameter L_1) has only slight effect on the contact characteristics.

In Fig. 3.16, graphs of the dimensionless external load as a function of the approach of the bodies are presented for the parameters of mutual influence of asperities $L_1 = 5$ and $L_1 = 50$ and the adhesion parameter $\lambda = 0.1$. For comparison, curves are shown with the same values of L_1, but without taking into account adhesion (*non-adhesive*) and the curve in the absence of roughness, i.e., Hertzian dependence (*Hertz*). It is assumed that $s = 2$, and the punch has a parabolic shape. The results indicate that the

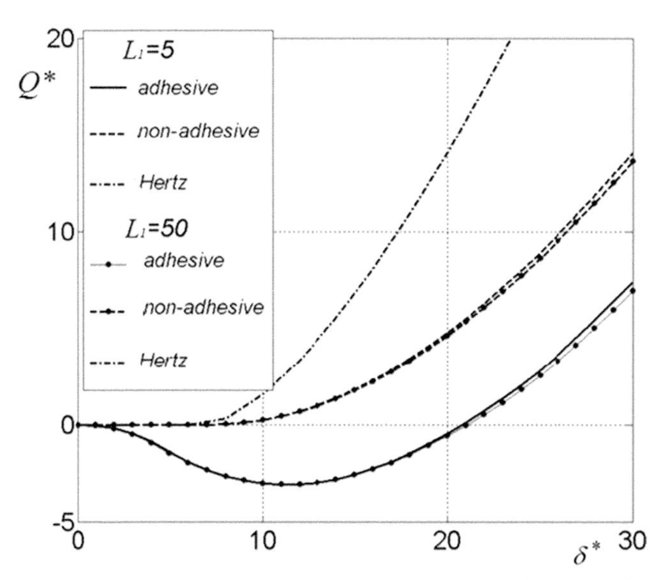

Figure 3.16 Load versus approach dependence at macrolevel for two different values of the parameter of mutual influence of contacts at microlevel.

presence of adhesion leads to a nonmonotonic dependence of the load on the distance between the bodies, while the rough adhesive contact is able to withstand negative loads.

3.5.4 Analysis of the real contact characteristics taking into account macro- and microgeometries of contacting bodies

The nominal pressure calculated from Eq. (3.38) or its particular form Eq. (3.48) can be used to determine the characteristics of a discrete contact such as real contact area, real pressure distribution, gap between the contacting bodies, and so on.

To examine the algorithm of this procedure, let us calculate the real contact area A_r in contact of the elastic cylinder, whose macroshape is described by the function $f(x) = x^2/(2R_0)$ (R_0 is the radius of the cylinder), and an elastic thick layer bonded with a rigid foundation, for various parameters characterizing their surface microgeometry. We consider the following models of the surface microgeometry: one- and three-level systems of spherical asperities uniformly distributed over the surface of the contacting bodies (see Chapter 1). The additional displacement functions $C(p)$ for these types of surface microgeometry are presented in Fig. 3.5. The dependence of the real contact area at one period on the nominal contact pressure can be calculated based on the method described in Chapter 1. Fig. 3.17 illustrates the variation of the relative real contact area $\lambda = \dfrac{4\pi\left(a_1^2 + a_2^2 + a_3^2\right)}{l^2\sqrt{3}}$ with the dimensionless nominal contact pressure $\dfrac{\pi\overline{p}}{2E^*}$ calculated for the one-level ($a_1 = a_2 = a_3$) and three-level models of asperity arrangement for the same parameters of surface microgeometry as in Fig. 3.5.

Taking into account the additional displacement function $C(p)$ (which depends on the surface microgeometry characteristics) and using the method of solving Eq. (3.38), we can calculate the nominal contact pressure $p(x, y)$ and the radius of the nominal contact region Ω at macroscale for the given macroshapes of contacting bodies and the load applied to them.

Then the real contact area A_r is determined from the expression

$$A_r = \iint_{\Omega} \lambda[p(x, y)]\,dxdy \tag{3.62}$$

and it depends both on macro- and microgeometries of contacting bodies.

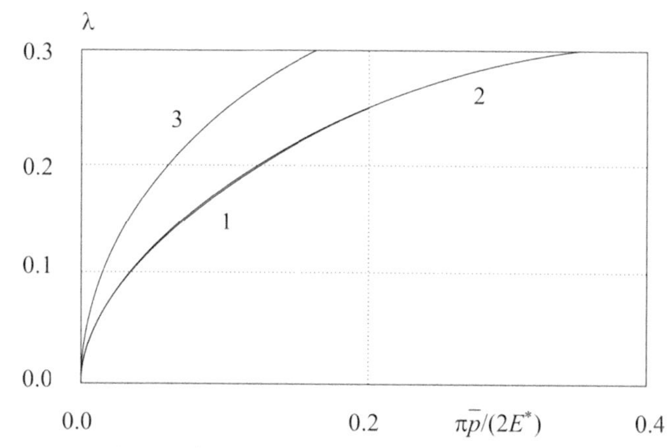

Figure 3.17 Dependence of the real area of contact on the nominal pressure for a three-level model (1) and for one-level models with $l/R = 0.6$ (2) and $l/R = 0.3$(3).

Fig. 3.18 illustrates the dependencies of the relative real contact area in contact of the cylinder and the layer (Fig. 3.12), which is determined by the formula

$$\frac{A_r}{A_a} = \int_{-1}^{1} \lambda(\overline{p}(\xi))d\xi \tag{3.63}$$

on the dimensionless load applied to the cylinder calculated for three models of surface microgeometries (one-level and three-level models with the asperity arrangement for the same parameters of surface

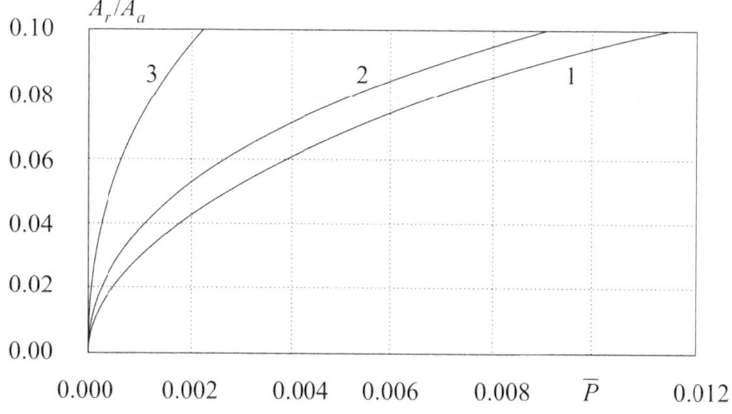

Figure 3.18 The dependence of the relative real contact area on the load applied to the cylinder for various microgeometry parameters.

microgeometry as in Fig. 3.5). The functions $C(p)$ and $\lambda(p)$ for these models are shown in Figs. 3.5 and 3.17, respectively.

A similar way can be used to calculate the maximum real contact pressure and the gap between the contacting bodies arising from their surface microgeometry. Estimation of the real contact pressure and its maximum values in contact of bodies with rough surfaces is of interest in studying the internal stresses in thin subsurface layers and wear of bodies in contact interaction. If the microgeometry of contacting bodies has a homogeneous structure along the surface, the maximum value of the real pressure occurs at contact spots where the nominal contact pressure reaches its peak. It can be calculated from the discrete contact problem solution (see Chapter 1) using the maximum values of the nominal pressure calculated for the contact problem at macroscale (see Section 3.5.2).

3.6 Conclusions

In this chapter, an approach to calculate the nominal contact characteristics in normal contact of elastic bodies of given macro shapes taking into account the parameters of their microgeometry (the asperities average shape, density of their location, and the height distribution) is presented. This approach is based on two-scale analysis:

- at microscale, the additional displacement due to surface microgeometry as a function of the nominal pressure is derived for a particular surface microgeometry of the contacting bodies, taking also into account the adhesive interaction in the gap between the contacting surfaces (if needed).
- at macroscale, the nominal contact characteristics are calculated taking into account the additional displacement function, which allows us to analyze the influence of microgeometry parameters as well as surface energy on the nominal contact characteristics.

The calculated nominal contact characteristics then can be used for analysis of the contact and internal stress distributions near the real contact spots taking into account the asperities curvature and spartial distribution.

It is worth noting that the method developed takes into account the mutual influence of contact spots, so it is applicable for high contact densities.

This approach allows us to analyze the influence of surface microgeometry on the contact characteristics at macroscale (the nominal contact area and nominal contact pressure distribution, and the approach of contacting bodies) and also to study the effect of macroshapes of the contacting bodies on the real contact area, gap between the contacting surfaces, and the maximum real contact pressure for given surface microgeometry parameters, their elastic properties, and surface energy.

References

Alexandrov, V.M., Kudish, I.I., 1979. Asymptotic analysis of plain and axisymmetric contact problems taking into account a surface structure of contacting bodies. Izv. AN SSSR Mech. Solids 1, 58–70.

Derjaguin, B.V., Muller, V.M., Toporov, Y.P., 1975. Effect of contact deformations on the adhesion of particles. J. Colloid Interface Sci. 53 (2), 314–326.

Galanov, B.A., 1984. Spatial contact problems for rough elastic bodies under elastoplastic deformations of asperities. J. Appl. Math. Mech. 48 (6), 1020–1029.

Galanov, B.A., 1985. The method of boundary equations of the Hammerstein-type for contact problems of the theory of elasticity with unknown contact regions. J. Appl. Math. Mech. 49 (5), 634–640.

Goryacheva, I.G., 1979. Plane and axisymmetric contact problems for rough elastic bodies. J. Appl. Math. Mech. 43 (1), 99–105.

Goryacheva, I.G., 2006. Mechanics of discrete contact. Tribol. Int. 39, 381–386.

Goryacheva, I.G., Makhovskaya, Y.Y., 2017a. Elastic contact between nominally plane surfaces in the presence of roughness and adhesion. Mech. Solid. 52 (4), 435–443.

Goryacheva, I.G., Makhovskaya, Y.Y., 2017b. Combined effect of surface microgeometry and adhesion in normal and sliding contacts of elastic bodies. Friction 05 (03), 339–350.

Goryacheva, I.G., Tsukanov, I.Y., 2018. Modelling of normal contact of elastic bodies with surface releif taken into account. J. Phys. Conf. 991, 12–28. IOP Publishing, Bristol, UK, England.

Greenwood, J.A., Tripp, J.H., 1967. The elastic contact of rough spheres. Trans. ASME, Ser. E. J. Appl. Mech. 34, 153.

Johnson, K.L., Kendall, K., Roberts, A.D., 1971. Surface energy and the contact of elastic solids. Proc. Roy. Soc. Lond. A. 324, 301–313.

Kragelsky, I.V., 1965. Friction and Wear. Butterworths, London.

Kragelsky, I.V., Dobychin, M.N., Kombalov, V.S., 1982. Friction and Wear: Calculation Methods. Pergamon.

Kuznetsov, E.A., 1978. Periodic contact problem accounting for the additional load outside the contact zone. Izv. Acad. Nauk SSSR Mekh.Tverd. Tela 6, 35–44 (in Russian).

Maugis, D., 1992. Adhesion of spheres: the JKR-DMT transition using a Dugdale model. J. Colloid Interface Sci. 150, 243–269.

Popov, G.Y., Savchuk, V.V., 1971. Contact problem for a circle contact region taking into account a surface structure of contacting bodies. Izv. AN SSSR Mech Solids 3, 80–87.

Rabinovich, A.S., 1975. Axisymmetric contact problems for rough elastic bodies. Mech. Solid. 4, 52–57 (In Russian).

Shtaierman, I.Y., 1949. Contact Problem of Theory of Elasticity. Moscow, Gostekhizdat (in Russian).

Sviridenok, A.I., Chizhik, S.A., Petrokovets, M.I., 1990. Mechanics of Discrete Frictional Contact. Minsk: Science and engineering (in Russian).

Teply, M.I., 1981. Problem of inner compression of cylindrical bodies with surface layer of higher compliance. Phys. & Chem. Mech. Mater. 2, 88–91 (In Russian).

Tsukanov, I.Y., 2017. Effects of shape and scale in mechanics of elastic interaction of regular wavy surfaces. Proc. IMech Part J J Eng. Tribol. 231 (3), 332–340.

Tsukanov, I.Y., 2018. Periodic contact problem for a surface with two-level waviness. Mech. Solid. 53 (Suppl. 1), 129–136.

Vorovich, I.I., Alexandrov, V.M., Babeshko, V.A., 1974. Nonclassical Mixed Problems of Elasticity Theory (IR) Nauka, Moscow, 456pp.

CHAPTER 4

Moving contact of elastic bodies with surface microgeometry

In this chapter, the process of normal approach and retraction of two asperities in the presence of adhesion is considered. The energy loss in cyclic interaction is calculated. The results obtained are used for modeling the friction force in sliding and rolling of rough elastic bodies.

4.1 Adhesive mechanism of energy dissipation in approach-retraction cycle of two individual asperities

Typical experimental dependencies of the normal load on the distance between two surfaces obtained with the use of an atomic force microscope (Jacquot and Takadoum, 2001) and adhesiometer (Grigoriev et al., 2003) have a hysteresis loop that cannot be completely explained by hysteresis mechanisms, such as surface roughness, imperfect elasticity, and chemical inhomogeneity.

The results of modeling confirm the possibility of hysteresis in adhesive contact of elastic bodies. Ambiguity of load-distance curves was established numerically (Greenwood, 1997) for two elastic spheres taking into account Lennard-Jones potential of molecular adhesion.

The area of the hysteresis loop, which is equal to the energy loss in an approach-retraction cycle, was calculated by Goryacheva and Makhovskaya (2001) for the cases of molecular adhesion described by Maugis model given by Eqs. (2.6)−(2.7) and capillary adhesion. The energy dissipation when bodies are moved into and out of contact was also evaluated by Wei and Zhao (2004) based on calculation of the energy release rate at the contact edge.

Discrete Contact Mechanics with Applications in Tribology
ISBN 978-0-12-821799-3
https://doi.org/10.1016/B978-0-12-821799-3.00003-0

In this section, we present the solution for two axisymmetric asperities in normal adhesive contact and, based on this solution, describe the adhesion hysteresis.

4.1.1 Contact problem solution for two asperities in the presence of adhesion

We consider the contact between two axisymmetric elastic asperities in the presence of capillary or molecular adhesion (Fig. 4.1). The gap between the surfaces before loading is described by a power function $f(r) = f_1(r) + f_2(r) = Ar^{2n}$, where n is an integer.

The solution to this problem follows from the results obtained in Chapter 2 for a periodic system of asperities. For this purpose, we set the period to infinity, $l \to \infty$. As a result, from Eqs. (2.23) and (2.24), we have the following relations for the distance between the interacting asperities d and normal load q:

$$\begin{aligned} d &= -\frac{(2n)!!}{(2n-1)!!}Aa^{2n} + \frac{2p_0 b}{E^*}\sqrt{1-\left(\frac{a}{b}\right)^2}, \\ q &= \frac{(2n)!!}{(2n+1)!!}4E^* Ana^{2n+1} - 2p_0 b^2\left(\arccos\frac{a}{b} + 2\frac{a}{b}\sqrt{1-\left(\frac{a}{b}\right)^2}\right), \end{aligned} \tag{4.1}$$

where $(E^*)^{-1} = (1-\nu_1^2)E_1^{-1} + (1-\nu_2^2)E_2^{-1}$ is the reduced elastic modulus of two interacting asperities. Eq. (4.1) defines the parametric dependence of the load q on distance d via the contact radius a as a parameter. To determine the unknown parameter b that is the outer radius of the meniscus (for capillary adhesion) or region of adhesive attraction (for molecular adhesion),

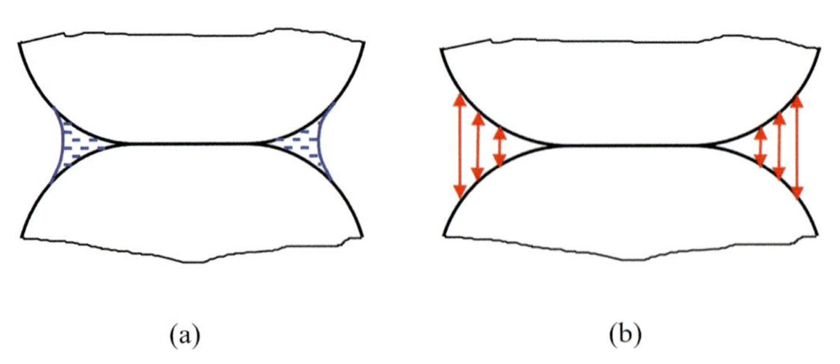

Figure 4.1 Contact of two asperities in the presence of capillary (a) and molecular (b) adhesion.

we have the following equation, which is obtained from Eq. (2.25) for the gap h between the asperities as $l \to \infty$ and from Eq. (2.12) or (2.14):

$$-\frac{2Aa^{2n}}{\pi}\left\{\arccos\frac{a}{b}\left[\left(\frac{b}{a}\right)^{2n}-\frac{(2n)!!}{(2n-1)!!}\right]\right.$$
$$\left.+\sqrt{\left(\frac{b}{a}\right)^2-1}\sum_{k=1}^{n}\frac{(2k-2)!!}{(2k-1)!!}\left(\frac{b}{a}\right)^{2(n-k)}\right\}p_0 \tag{4.2}$$
$$+\frac{4b}{\pi E^*}\left\{1-\frac{a}{b}-\arccos\frac{a}{b}\sqrt{1-\left(\frac{a}{b}\right)^2}\right\}p_0^2+2\gamma_0=0.$$

Here, γ_0 is the surface tension of fluid for the case of capillary adhesion, and $\gamma_0 = w_a/2$ (w_a is the specific work of adhesion) for the case of molecular adhesion.

In the case of capillary adhesion, we need one more equation for the determination of the unknown adhesive pressure p_0, which is the Laplace pressure in the meniscus. This equation follows from the condition of volume conservation given by Eq. (2.26) for $l \to \infty$:

$$v_0 = 2Aa^{2n+2}\left\{\sqrt{\left(\frac{b}{a}\right)^2-1}\left[\frac{(2n)!!(2n-1)}{(2n+1)!!}\right.\right.$$
$$\left.+\frac{1}{n+1}\sum_{k=0}^{n}\frac{(2k)!!}{(2k+1)!!}\left(\frac{b}{a}\right)^{2(n-k)}\right]$$
$$\left.-\left(\frac{b}{a}\right)^2\left[\frac{(2n)!!}{(2n-1)!!}-\left(\frac{b}{a}\right)^{2n}\frac{1}{n+1}\right]\arccos\frac{a}{b}\right\} \tag{4.3}$$
$$-\frac{4p_0b^3}{3E^*}\left[4-3\frac{a}{b}-\left(\frac{a}{b}\right)^3-3\arccos\frac{a}{b}\sqrt{1-\left(\frac{a}{b}\right)^2}\right]$$

In the case of no contact between the two interacting asperities, from Eq. (2.27) for $l \to \infty$, we obtain the following relations for the distance between the interacting asperities d and normal load q:

$$q = -\pi p_0 b^2, \quad d = -Ab^{2n} + \frac{4p_0 b}{\pi E^*} + \frac{\gamma_0}{p_0}. \tag{4.4}$$

Eq. (4.4) defines the parametric dependence of the load q on the distance d via the parameter b. In the case of capillary adhesion, from Eq. (2.28), we have the equation for the determination of the Laplace pressure p_0:

$$v_0 = \pi b^2 \left(\frac{Ab^{2n}}{n+1} + d \right) - \frac{16 p_0 b^3}{3E^*}. \tag{4.5}$$

The relations presented make it possible to analyze the dependence of the external normal load q on the distance d between the two asperities in a quasistatic process of normal approach and retraction.

4.1.2 Analysis of the load-distance dependence

In Fig. 4.2, load-distance curves are presented for two asperities whose shape is described by the function $f(r) = Ar^{2n}$, $n = 2$. Thick lines correspond to the contact between the asperities and thin lines to the contactless interaction.

The plots presented in Fig. 4.2a are constructed for the case of molecular adhesion described by Eqs. (2.6)–(2.7) with using Eqs. (4.1)–(4.2) for the contact and Eq. (4.4) for the contactless interaction. In this figure, the dimensionless load $q/(p_0 D^2)$ (where $D = (2A)^{-1/(2n-1)}$) is shown as a function of the dimensionless distance $\delta = d/D$ between the asperities for $E^*/p_0 = 1$. Curves 1 and 2 are constructed for two values of the dimensionless specific work of adhesion, $w_a/(p_0 D) = 1$ and $w_a/(p_0 D) = 2$, respectively.

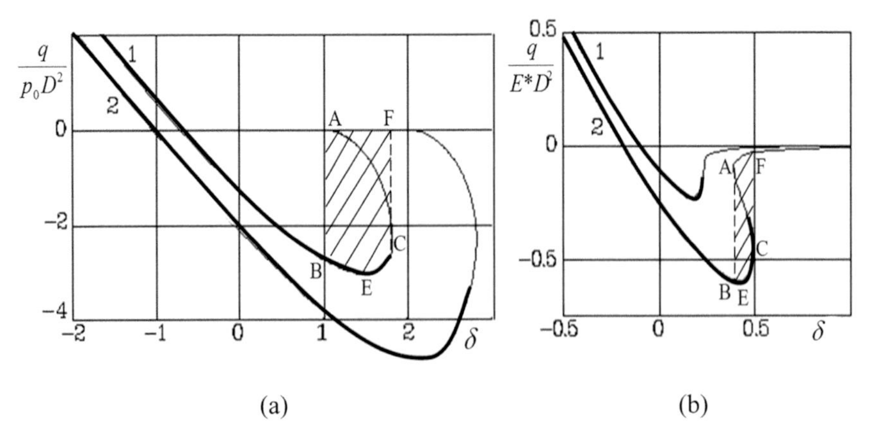

Figure 4.2 Dimensionless load as a function of distance for the cases of molecular (a) and capillary (b) adhesion.

In Fig. 4.2b, the dimensionless load $q/\left(E^{*}D^{2}\right)$ versus distance $\delta = d/D$ is presented for the case of capillary adhesion with the meniscus volume $v_0/D^3 = 0.05$. The results are calculated by Eqs. (4.1)–(4.2) for the contact and by Eqs. (4.4) and (4.5) for the contactless interaction. Curve 1 corresponds to the dimensionless surface tension of fluid in the meniscus $\gamma_0/\left(E^{*}D\right) = 0.025$, and curve 2 corresponds to $\gamma_0/\left(E^{*}D\right) = 0.05$.

The results indicate that for both molecular and capillary adhesion, the load-distance curves are ambiguous for high enough values of the dimensionless specific work of adhesion $w_a/(p_0 D)$ and surface tension of fluid $\gamma_0/\left(E^{*}D\right)$, respectively.

The dependencies presented in Fig. 4.2 allow us to analyze the processes of normal approach and retraction of two asperities in the presence of adhesion. Consider curve 1 in Fig. 4.2a and curve 2 in Fig. 4.2b. If the asperities are moved away from each other under the controlled (monotonically decreased) load q, then as the load attains its minimum value $q_{\min}$ (point E), the jump breaking of contact occurs in the cases of both molecular and capillary adhesion. The force $q_{\min}$ of this jump is called the pull-off force. If the asperities are moved away from each other under the controlled (monotonically increased) distance d between them, then the jump occurs from point C to point F. If the asperities are moved toward each other, as the distance between them is decreased, they jump from point A to point B. Note that points A and F always correspond to the contactless interaction, whereas points B and C can correspond to either contact or contactless cases, depending on the values of the dimensionless specific work of adhesion $w_a/(p_0 D)$ or surface tension of fluid $\gamma_0/\left(E^{*}D\right)$ and its volume v_0/D^3. Therefore, due to adhesion the asperities can jump into contact and out of contact.

This process is illustrated in Fig. 4.3 for the capillary adhesion. It is assumed that in the case of capillary adhesion, the cyclic process of approach

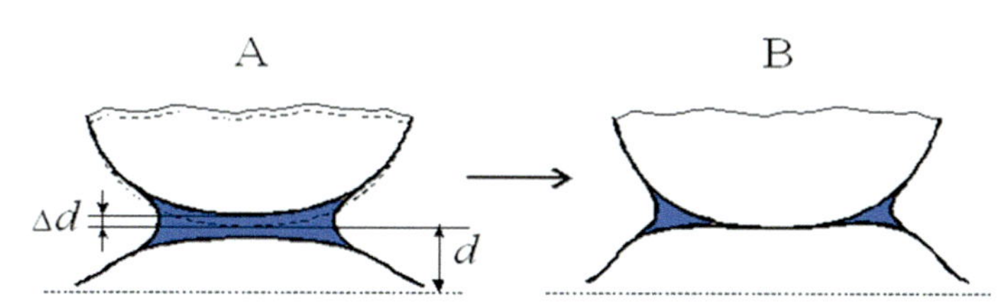

Figure 4.3 Scheme of two asperities jumping into contact in the presence of capillary adhesion.

and retraction of asperities, including their jumps into and out of contact, occurs without breaking the meniscus. Besides, the assumptions under which the model of capillary adhesion was constructed (Section 2.2) include the restrictions on the shape of the meniscus; in particular, its diameter (width) should be considerably larger than its height. So, this model cannot be used to describe the contactless interaction of two asperities in the presence of a meniscus, as they are relatively far from each other. Also, this model does not describe breaking of the meniscus when the distance between the asperities becomes too large. Models taking into account possibility of a meniscus breaking at large distances but neglecting deformations caused by the meniscus were proposed by Chizhik (1994) and Chekina (1998).

Thus, from the analysis of the processes of approach and retraction of two asperities, it follows that the work done by the adhesion force when the surfaces move toward each other is lower than the work done by the external force when the surfaces are moved from each other. The corresponding energy loss is defined by the dashed areas in Fig. 4.2 and can be calculated from the relation

$$\Delta w = \int_{ABCF} q(d)\mathrm{d}d. \tag{4.6}$$

Based on the load-distance dependences for two axisymmetric asperities obtained in Section 4.1, the energy loss Δw and pull-off force $q_{\min}$ can be calculated for the cases of molecular and capillary adhesion.

Plots of the dimensionless energy loss $\Delta w/(p_0 D^3)$ and pull-off force $q_{\min}/(p_0 D^2)$ versus the dimensionless specific work of adhesion $w_a/\ (p_0 D)$ are presented in Fig. 4.4a and b, respectively, for the case of molecular adhesion of dry surfaces. Solid lines correspond to $n = 1$, i.e., the asperities having the shape of paraboloids of revolution. The results indicate that as the specific work of adhesion of two surfaces increases, the energy loss in their cycling approach and retraction also increases and then attains a constant value. The energy loss $\Delta w/p_0 D^3$ is higher for lower values of E^*/p_0, i.e., for softer asperities. The pull-off force $q_{\min}/(p_0 D^2)$, on the contrary, increases with increasing E^*/p_0. For comparison, plots for a different shape of the interacting asperities ($n = 2$) and $E^*/p_0 = 2$ are presented by dashed lines.

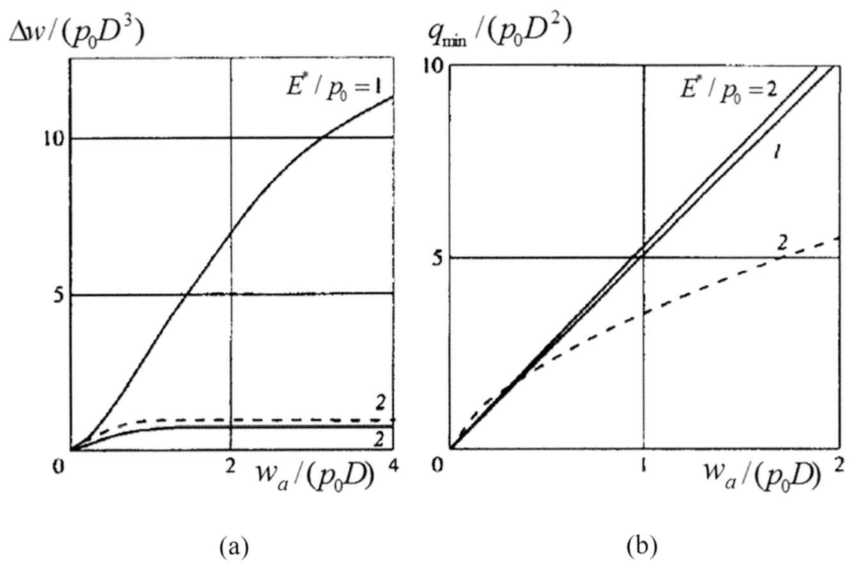

Figure 4.4 Dimensionless energy loss (a) and pull-off force (b) versus dimensionless specific work of adhesion in the case of molecular adhesion.

In Fig. 4.5, plots of the dimensionless energy loss $\Delta w/(p_0D^3)$ versus the dimensionless surface tension of fluid in the meniscus $\gamma_0/(E^*D)$ are presented for the case of capillary adhesion ($n = 1$, solid lines). The analysis of the problem solution shows that the energy loss differs from zero only

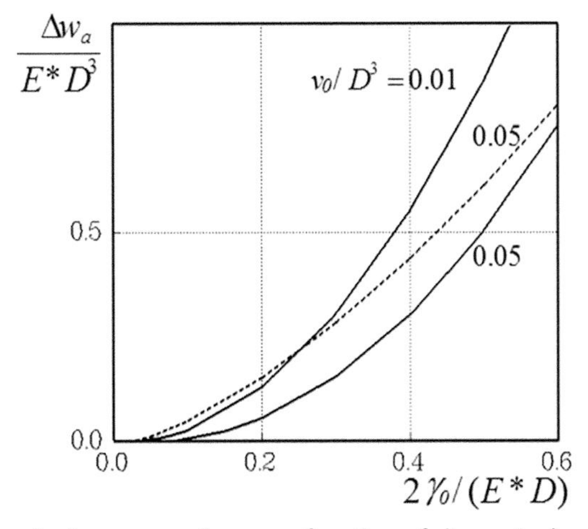

Figure 4.5 Dimensionless energy loss as a function of dimensionless surface tension of fluid in the meniscus for the case of capillary adhesion.

beginning from a certain value of the dimensionless surface tension, and it increases without limit as this parameter increases. The value of $\Delta w/p_0 D^3$ is higher for greater volumes of the meniscus. A plot for a different shape of asperities ($n = 2$) and $v/D^3 = 0.05$ is also presented (dashed line) in Fig. 4.5.

4.1.3 Energy dissipation in approach and retraction of paraboloids of revolution

The case $n = 1$ corresponds to the interaction of two asperities whose shape is described by the parabolic function $f(r) = Ar^2$. In this case, the quantity $D = (2A)^{-1/(2n-1)}$ is the reduced radius of curvature of their tops: $D = R_1^{-1} + R_2^{-1} = R$.

Let us consider the case of molecular adhesion described by the Maugis model given by Eqs. (2.6) and (2.7). The analysis of Eqs. (4.1), (4.2), and (4.4) shows that the dependence of the dimensionless load Q_1 on the dimensionless distance δ_1, which is defined by the expressions

$$Q_1 = \frac{q}{\pi R w_a}, \quad \delta_1 = d\left(\frac{16E^{*2}}{9\pi^2 w_a^2 R}\right)^{1/3}, \tag{4.7}$$

is described by the only parameter λ introduced by Eq. (3.29). This parametrization was first used by Maugis (1992) for the adhesive contact between a paraboloid of revolution and an elastic half-space. The adhesion parameter is similar and close to the Tabor parameter given by Eq. (2.5).

Plots of the dimensionless force Q_1 as a function of the dimensionless distance δ_1 are presented in Fig. 4.6a for $\lambda = 0.1$, 0.6, and 2 (curves 1, 2, and 3, respectively). Thick lines correspond to the contact of asperities and thin lines to their contactless interaction. In the limit case $\lambda \to 0$, the load-distance curve coincides with the DMT simplified model of adhesion. In the case of $\lambda \to \infty$, the solution tends to the JKR simplified model. The results of calculation indicate that for small λ, the dependence of Q_1 on δ_1 is virtually unambiguous (see curve 1 in Fig. 4.6a). As the adhesion parameter increases, the load-distance curve becomes more ambiguous, and accordingly, the energy loss in an approach-retraction cycle increases with increasing λ.

This conclusion is supported by the calculation of the dimensionless energy loss in an approach-retraction cycle

$$\Delta W = \Delta w \left(\frac{16E^{*2}}{9\pi^5 w_a^5 R^4}\right)^{1/3} \tag{4.8}$$

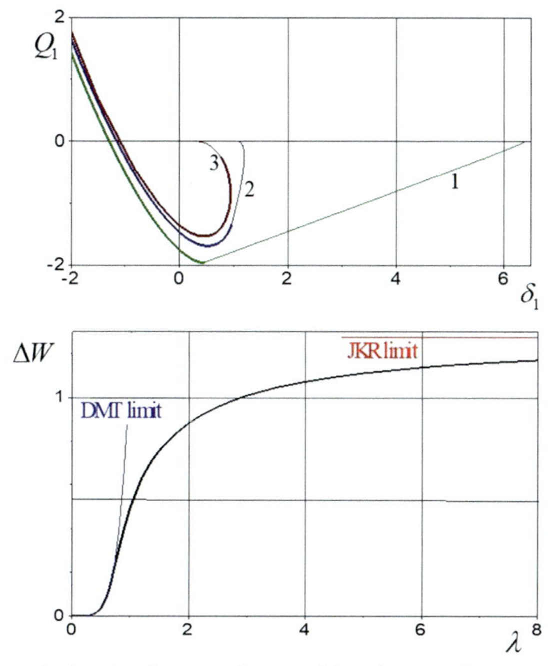

Figure 4.6 Dimensionless load versus distance (a) and energy loss versus the adhesion parameter (b) in the case of molecular adhesion between two parabolic asperities.

as a function of the adhesion parameter λ. The plot of $\Delta W(\lambda)$ is presented in Fig. 4.6b. As $\lambda \to 0$, the dimensionless energy loss ΔW tends to zero. For $0 \le \lambda \le (9/32)^{1/3}$, the hysteresis loop lies entirely in the contactless region (thin lines). In this case, the load-distance dependence is defined by Eq. (4.4), which with the parametrization Eq. (4.7) leads to the following:

$$\delta_1 = \frac{8\lambda}{3\pi}\beta_1 - \frac{\beta_1^2}{2} + \frac{2}{\pi\lambda}, \; Q_1 = -\frac{1}{2}\pi\lambda\beta_1^2, \quad \text{where } \beta_1 = b\left(\frac{4E^*}{3\pi w_a R^2}\right)^{1/3}. \tag{4.9}$$

The energy loss is then calculated as follows:

$$\Delta W_{\mathrm{DMT}} = \int\limits_{\beta_1^*}^{\beta_1^{**}} Q_1(\beta)\delta_1'(\beta)d\beta, \quad \text{where } \beta_1^* = \frac{8\lambda}{3\pi}, \; \beta_1^{**} = \frac{16\lambda}{3\pi}, \tag{4.10}$$

where the functions $Q_1(\beta_1)$ and $\delta_1(\beta_1)$ are defined by Eq. (4.9). After taking the integral in Eq. (4.10), we obtain the following expression for the function $\Delta W(\lambda)$:

$$\Delta W_{\mathrm{DMT}} = \frac{8704}{243\pi^3}\lambda^5, \quad 0 \le \lambda \le \left(\frac{9}{32}\right)^{1/3}. \tag{4.11}$$

For $\lambda \to \infty$, the hysteresis loop in the load-distance dependence lies completely within the contact domain (thick lines in Fig. 4.6). In this limit case, from Eqs. (4.1) and (4.2) with parametrization Eq. (4.7), we have the following relations:

$$\delta_1 = -\alpha_1^2 + \frac{2}{3}\sqrt{6\alpha_1}, \quad Q_1 = \alpha_1^3 - \alpha_1\sqrt{6\alpha_1}, \quad \text{where } \alpha_1 = a\left(\frac{4E^*}{3\pi w_a R^2}\right)^{1/3}, \tag{4.12}$$

which coincide with the JKR equations. Then the energy loss calculated from the formula

$$\Delta W_{JKR} = \int_{\alpha_1^*}^{\alpha_1^{**}} Q_1(\alpha)\delta_1'(\alpha)d\alpha, \quad \text{where } \alpha_1^* = \frac{1}{6^{1/3}}, \alpha_1^{**} = \frac{2}{3^{1/3}}, \tag{4.13}$$

has the following constant value:

$$\Delta W_{JKR} = \frac{3^{1/3}\left(2^{1/3} + 12\right)}{15} \approx 1.28 \quad \text{for } \lambda \to \infty. \tag{4.14}$$

In the case of capillary adhesion, the dependence of the dimensionless load Q_1 on the distance δ_1 specified by Eq. (4.7), in which we set $w_a = 2\gamma_0$, is also defined by the only dimensionless parameter:

$$\eta = \frac{1}{\nu_0}\left(\frac{\gamma_0^4 R^5}{E^{*4}}\right)^{1/3} \tag{4.15}$$

The capillary adhesion parameter η defined by Eq. (4.15) is the analog of the molecular adhesion parameter λ. The dimensionless energy loss given by Eq. (4.8) in the case of capillary adhesion is the function of the only parameter η. The plot of the function $\Delta W(\eta)$ is presented in Fig. 4.7b. It is seen that the value of ΔW differs from zero only for $\eta \ge \eta_0$ ($\eta_0 \approx 1.76$).

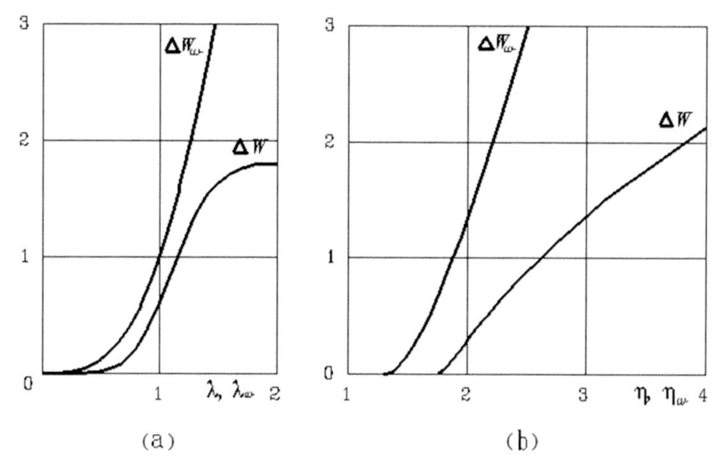

Figure 4.7 Dimensionless energy loss for the half-space and Winkler models versus adhesion parameter in the cases of molecular (a) and capillary (b) adhesion.

Unlike the case of molecular adhesion, for capillary adhesion the energy loss unlimitedly grows as the parameter η increases.

4.1.4 Adhesive contact problem solution using the Winkler model

The solution of the problem of adhesive contact between two elastic axisymmetric asperities obtained in Section 4.1.1 based on the elastic half-space model requires numerical solution of Eq. (4.2) for molecular adhesion or the system of Eqs. (4.2) and (4.3) for capillary adhesion to determine the dependence of the load upon distance Eq. (4.1). This solution can be significantly simplified if we assume that the elastic properties of the interacting asperities are described by the Winkler model (spring foundation). In this case, Eq. (2.19), which relates normal displacements to pressures at the boundary of the elastic half-space, is replaced by a simple dependence:

$$u_z(r) = kp(r), \quad k = k_1 + k_2 \tag{4.16}$$

where k_1 and k_2 are the spring constants of two interacting asperities. For simplicity we assume that they have the shape of paraboloids of revolution, i.e., $n = 1$, and let R be the reduced radius of the tops of the interacting asperities: $R^{-1} = R_1^{-1} + R_2^{-1}$. In this case, the value of gap between the surfaces in the region of action of the adhesive pressure p_0 is defined by the relation:

$$h(r) = \frac{r^2}{2R} + d - kp_0. \tag{4.17}$$

For molecular adhesion of dry surfaces, taking into account the Winkler model relations Eqs. (4.16) and (4.17), we obtain the following solution:

in the case of contact between the asperities, for $h(0) = 0$,

$$q = -2\pi R w_a + \frac{\pi R}{k}\left(d^2 - k^2 p_0^2\right), \tag{4.18}$$

and in the contactless case, for $h(0)>0$,

$$q = -2\pi R w_a + 2\pi R p_0 (d - k p_0). \tag{4.19}$$

The obtained solution for the Winkler model defines a simple dependence of the load q on the distance d between the asperities in the closed form. If we introduce the dimensionless load Q_1 in accordance with the first relation of Eq. (4.7) and dimensionless distance $\delta_w = d(kw_a)^{-1/2}$, then the load-distance dependence contains only one dimensionless parameter.

In the case of molecular adhesion, this parameter is

$$\lambda_w = p_0 \sqrt{k/w_a}, \tag{4.20}$$

by using which we represent Eqs. (4.18) and (4.19) in the form

$$Q_1 = \begin{cases} \delta_w^2 - 2 - \lambda_w^2, & \delta_w \le \lambda_w \\ 2\left(\lambda_w \delta_w - \lambda_w^2 - 1\right), & \delta_w > \lambda_w \end{cases}. \tag{4.21}$$

For $\delta_w \le \lambda_w$, the asperities are in contact, whereas for $\delta_w > \lambda_w$, their interaction is contactless. The case $\delta_w = \lambda_w$, $Q_1 = -2$ corresponds to the point contact.

In the case of capillary adhesion, apart from Eqs. (4.18) and (4.19), we should use the condition of conservation of the volume of fluid in the meniscus. By substituting the expression for the gap given by Eq. (4.17) into the equation for the fluid volume, Eq. (2.15), we obtain

$$v_0 = \pi\left(b^2 - a^2\right)\left(\frac{a^2 + b^2}{4R} + d - kp_0\right). \tag{4.22}$$

In the contactless case, one should set $a = 0$ in this expression. After this, by passing to the dimensionless quantities Q_1 and δ_w, and introducing the parameter

$$\eta_w = \frac{2\pi R k \gamma_0}{v_0}, \tag{4.23}$$

we obtain

$$\delta_w^2 = \begin{cases} Q_1 + 2 + \eta_w, & \delta_w \le \eta_w \\ -\dfrac{1}{\eta_w Q_1} \dfrac{\left(2 + Q_1 - 2Q_1\eta_w - Q_1^2\eta_w/2\right)^2}{Q_1 + 4}, & \delta_w > \eta_w \end{cases}. \tag{4.24}$$

For $\delta_w \le \eta_w$ the asperities are in direct contact, and for $\delta_w > \eta_w$, they are in contactless interaction through the meniscus. The case $\delta_w = \eta_w$, $Q_1 = -2$ corresponds to the point contact of the asperities.

From Eqs. (4.22) and (4.24), it follows that the dependence of the load Q_1 applied to the asperities on the distance δ_w between them is non-monotone for both capillary and molecular adhesion. This dependence is unambiguous in the case of molecular adhesion and ambiguous for the capillary one. The comparison of the relations obtained with the solution for the elastic half-space (described by Eq. (2.19) between the elastic stress and displacement) shows that the Winkler model gives qualitatively agreeable result for the case of capillary adhesion. In the case of molecular adhesion, qualitative agreement takes place only for the case of contact between the surfaces.

Based on Eqs. (4.22) and (4.24), we determine the dimensionless energy loss in an approach-retraction cycle of two asperities

$$\Delta W_w = \frac{\Delta w}{\pi R} \sqrt{\frac{1}{k w_a^3}} \tag{4.25}$$

as a function of the parameters λ_w and η_w for the cases of molecular and capillary adhesion, respectively (in the case of capillary adhesion, the quantity w_a should be replaced by $2\gamma_0$). These dependencies are presented in Fig. 4.7. In the case of molecular adhesion, the function $\Delta W_w(\lambda_w)$ is obtained analytically:

$$\Delta W_w = \begin{cases} \lambda_w^3, & \lambda_w \le 1 \\ \dfrac{1}{3} \dfrac{\left(\lambda_w^2 - 1\right)\left(2\lambda_w^4 + 5\lambda_w^2 - 1\right)}{\lambda_w^3} + \dfrac{1}{\lambda_w}, & \lambda_w > 1 \end{cases}. \tag{4.26}$$

As it is seen in Fig. 4.7, this function is close to the corresponding function for the elastic half-space only for small values of the parameter λ_w.

In the case of capillary adhesion, the plot of the function $\Delta W_w(\eta_w)$ for the Winkler foundation is qualitatively similar to the function $\Delta W(\eta)$ for the elastic half-space. The value η_{w0} starting from which the energy loss ΔW_w differs from zero is determined analytically $\eta_{w0} = 3\sqrt{2}/4 \approx 1.30$.

4.1.5 Effect of the asperity shape on the energy loss during an approach-retraction cycle

In the case of $n > 1$ (nonparabolic asperities), two dimensionless parameters are necessary to describe the load-distance dependence for the adhesive interaction. We consider the case of molecular adhesion and choose the following two parameters. The first one, E^*/p_0, is the ratio of the reduced elastic modulus of two asperities to the adhesion pressure in the Maugis model given by Eqs. (2.6) and (2.7). The second parameter, $w_a/(p_0 D)$, is the ratio of the radius $h_0 = w_a/p_0$ of adhesion force to the characteristic size of the asperities $D = (2A)^{-1/(2n-1)}$.

The solution for two asperities presented in Section 4.1.1 allows us to analyze the effect of their shape on the load-distance dependence. Examples of such dependencies are presented in Fig. 4.8a in which the plots of the dimensionless load $q/p_0 D^2$ are constructed as a function of the dimensionless distance $\delta = d/D$ for $E^*/p_0 = 1$ and $w_a/(p_0 D) = 10^{-6}$. Curves 1, 2, and 3 correspond to the asperities of different shape: $n = 1$, 2, and 3. The corresponding functions of shape, $f(r) = f_1(r) + f_2(r) = Ar^{2n}$, are shown in Fig. 4.8b for $n = 1$, 2, and 3 (curves 1, 2, and 3, respectively).

The results indicate that the shape of the asperities considerably influences the load-distance dependence. As the number n increases (tops of the asperities become flatter), the hysteresis loop area in the load-distance plot increases, and hence, the energy loss in an approach-retraction cycle increases.

Plots of the dimensionless energy loss $\Delta w/(p_0 D^3)$ versus the parameter $w_a/(p_0 D)$ are presented in Fig. 4.9 for $E^*/p_0 = 2$ and various asperities shapes. Curves 1, 2, 3 correspond to $n = 1$, 2, 3, respectively. It is apparent that as the parameter $w_a/(p_0 D)$ increases, the dimensionless energy loss increases and attains a constant value. This value can be estimated analytically. By substituting the solution Eq. (4.4) into Eq. (4.6) and taking the integral, we obtain

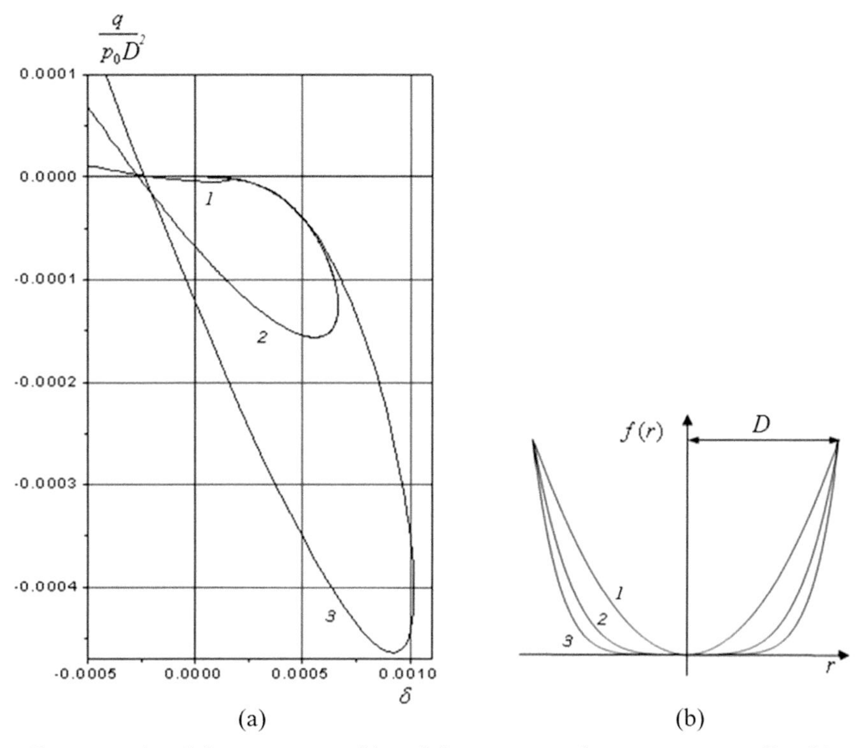

Figure 4.8 Load-distance curves (a) and the corresponding asperities profiles (b).

$$\frac{\Delta w}{p_0 D^3} = \frac{n\pi}{2n+2}\left(\xi_1^{2n+2} - \xi_2^{2n+2}\right) - \frac{4p_0}{3E^*}\left(\xi_1^3 - \xi_2^3\right) \text{ for } \frac{w_a}{p_0 D} \geq \frac{1}{2}\xi_1^{2n} + \frac{2(\pi-2)p_0}{\pi E^*}\xi_1, \tag{4.27}$$

$$\text{where } \xi_1 = \left(\frac{8p_0}{\pi E^*}\right)^{\frac{1}{2n-1}}, \quad \xi_2 = \left(\frac{4p_0}{n\pi E^*}\right)^{\frac{1}{2n-1}}$$

In the case $n = 1$, Eq. (4.27) coincides with Eq. (4.11).

4.2 Modeling of sliding friction force

The adhesion hysteresis in a normal approach-retraction cycle was measured experimentally and compared to the sliding friction force for the same pairs of surfaces (Chaudhury and Owen, 1993; Yoshizawa et al., 1993; Szoszkiewicz et al., 2005). It was established that the values of hysteresis and

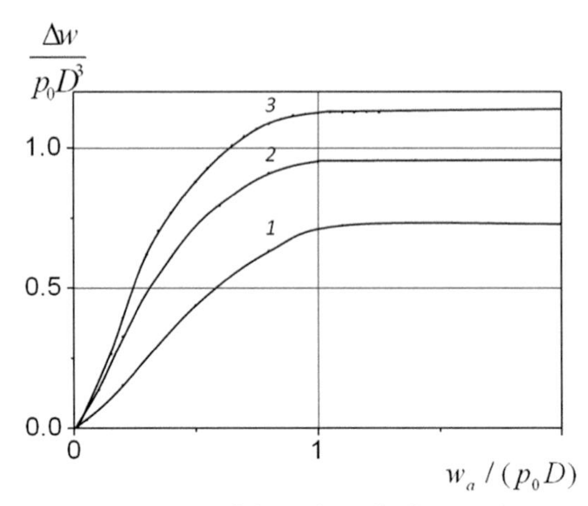

Figure 4.9 Energy loss versus ratio of the radius of adhesion force to the characteristic size of the asperities.

friction force correlate, so one can conclude that adhesion hysteresis gives an important contribution into the adhesive component of the friction force.

To model this mechanism of the adhesive friction force, the moving contact was represented as two cracks—opening and closing ones—with calculating the energy lost for fracture (Johnson, 1997). Within the framework of this approach, models were suggested to relate the adhesive friction force with the adhesion hysteresis of elastic bodies for a sliding cylinder (Barquins, 1990) and for sliding of a periodic wavy surface (Carbone and Mangialardi, 2004). In these models, hysteresis was taken into account by assuming different surface energy before and after the moving contact area. For the case of viscoelastic foundation, this approach was implemented by Goryacheva et al. (2014).

Another approach to taking the adhesion hysteresis into account was used (Heise and Popov, 2010; Goryacheva and Makhovskaya, 2011, 2017) based on the calculation of the energy dissipated in formation and breaking of adhesive contacts. It is substantial in this approach that both surfaces have surface relief, so that cyclic formation and breaking of elementary contacts occur during sliding. The model based on this approach is presented below to describe the adhesive component of the friction force in mutual sliding of two elastic bodies whose surfaces have regular relief. The model is constructed for dry surfaces interacting via molecular adhesion forces.

4.2.1 Mutual displacement of two asperities in sliding

Consider mutual displacement of two initially hemispherical asperities during sliding (Fig. 4.10). It is assumed that the lower asperity of radius R_1 is at rest, while the upper asperity of radius R_2 moves along the x-axis, with the distance between the asperities in the z-direction remaining constant. At first, the asperities do not interact with each other (Fig. 4.10a); then they come into contact and mutual sliding occurs (Fig. 4.10b) up to the moment when they come out of the contact with a jump (Fig. 4.10c). Thus, an approach-retraction cycle takes place in the tangential direction, which should be associated with an energy loss.

Since the asperities have a spherical shape, at each time instant the force of interaction between them acts along the axis O_1O_2 connecting the centers of the spheres (Fig. 4.11). The tangential stresses are assumed to be zero, so the contact problem at each time instant is symmetric with respect to the axis O_1O_2. The force q acting on the upper asperity from the lower one is defined by Eqs. (4.1)–(4.2) for the contact and by Eq. (4.4) for the contactless case. In these equations, we should set $n = 1$ (the tops of the spherical asperities are approximated by paraboloids of revolution).

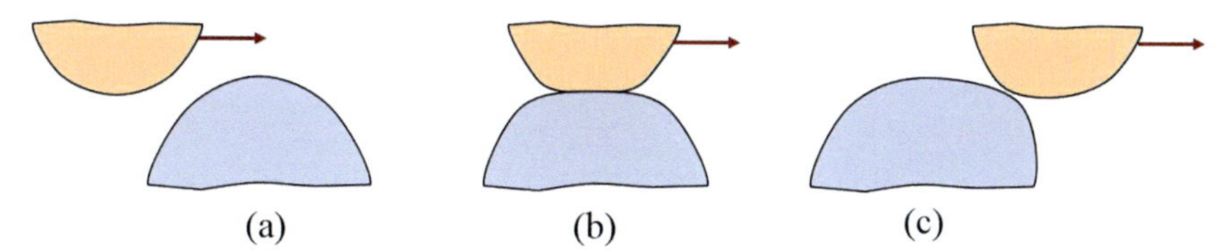

Figure 4.10 Scheme of mutual displacement of elastic asperities in sliding.

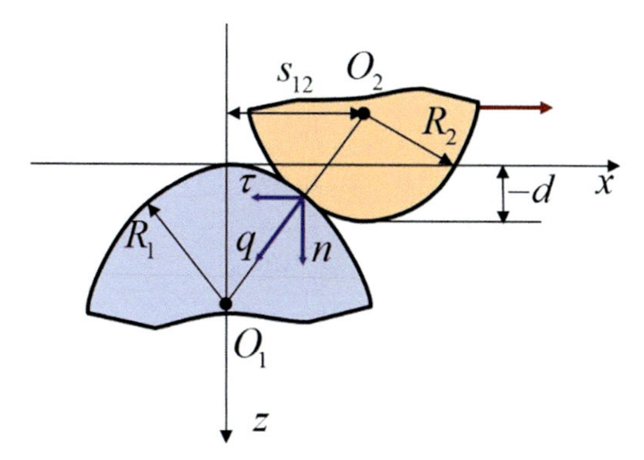

Figure 4.11 Scheme of force interaction between two asperities in sliding.

The distance δ between the asperities in the direction of the axis O_1O_2 is related to the distance s_{12} in the x-direction between the centers of the spheres and distance d in the z-direction by the formula $\delta = \sqrt{(R_1 + R_2 + d)^2 + s_{12}^2} - R_1 - R_2$. The force of interaction q has a normal n (along the z-axis) and tangential τ (along the x-axis) components:

$$n = \frac{q(R_1 + R_2 + d)}{\sqrt{(R_1 + R_2 + d)^2 + x^2}}, \qquad \tau = \frac{qx}{\sqrt{(R_1 + R_2 + d)^2 + x^2}}. \tag{4.28}$$

Plots of the dimensionless normal n and tangential τ forces acting on the upper asperity as it slides with respect to the lower asperity along the x-coordinate are shown in Fig. 4.12. The forces are calculated for the following values of the dimensionless parameters: $\frac{w_a}{p_0 R} = 10^{-3}$, $p_0/E^* = 1$, and $(R_1 + R_2)/R = 4$, where $\frac{1}{R} = \frac{1}{R_1} + \frac{1}{R_2}$. Curves 1 correspond to $d/R = -5 \times 10^{-4}$, and curves 2 to $d/R = -0.023$.

To calculate the force of sliding friction, we consider the tangential component τ with which the lower asperity acts on the upper one. During sliding of the upper asperity, the force τ changes its sign from positive (as it acts in the direction of motion) to negative (as it prevents sliding of the asperity). As a result of the hysteresis occurring in an approach-retraction cycle, the total work of this force calculated as

$$A_f = \int_{-\infty}^{\infty} \tau(s)\,ds \tag{4.29}$$

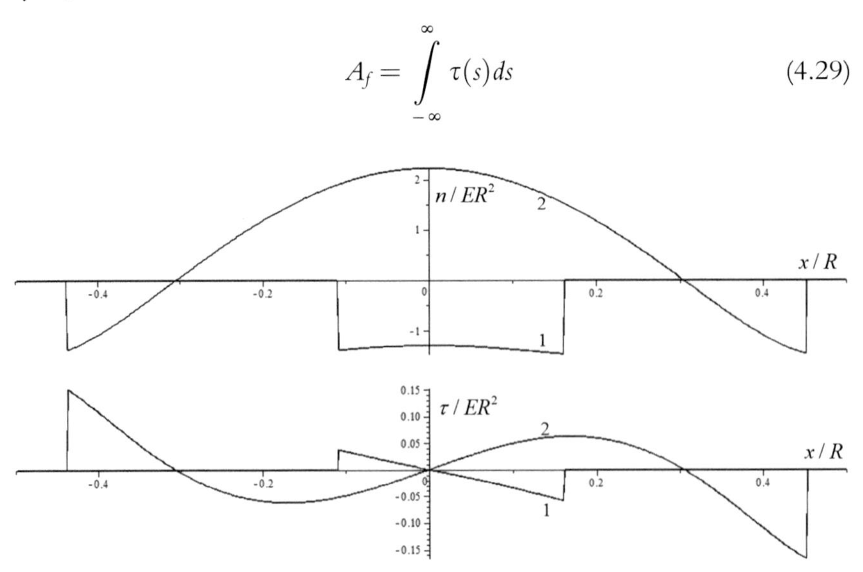

Figure 4.12 Normal and tangential components of the force of interaction between two asperities in sliding.

differs from zero. This work is equal to the energy lost in an approach-retraction cycle of two asperities, i.e., $A_f = \Delta w$.

In what follows, we consider an example of two surfaces with regular relief and calculate the friction force between them in sliding.

4.2.2 Sliding of two surfaces with regular relief

To simplify the calculation we assume that the upper and lower surfaces are both regular with the same period l (Fig. 4.13). When each asperity of the upper surface passes over a periodic segment of the lower surface, the tangential force acting on this asperity produces the work $A_f = \Delta w$ so that the average tangential force acting on this asperity is $\Delta w / l$.

Since each asperity occupies the area l^2, we have the following expression for the average tangential force acting on the upper surface:

$$\bar{\tau} = \Delta w / l^3. \tag{4.30}$$

As it was shown in Section 4.1.2, the energy loss Δw in an elementary cycle of normal approach-retraction of two asperities of paraboloidal shape can be represented in the dimensionless form as a function of the only parameter, $\Delta W(\lambda)$, where ΔW and λ are given by Eq. (4.8) and Eq. (3.29), respectively. The average tangential force acting on the upper regular surface, which is equal to the adhesive friction force, can also be represented in the dimensionless form

$$\overline{T} = \bar{\tau} l^3 \left(\frac{16E^{*2}}{9\pi^5 w_a^5 R^4} \right)^{1/3}. \tag{4.31}$$

From Eqs. (4.30) and (4.31), it follows that

$$\overline{T} = \Delta W(\lambda). \tag{4.32}$$

The graph of the function $\Delta W(\lambda)$ is presented in Fig. 4.6b. Thus, the dimensionless friction force $\overline{T}$ is determined for any value of the parameter λ.

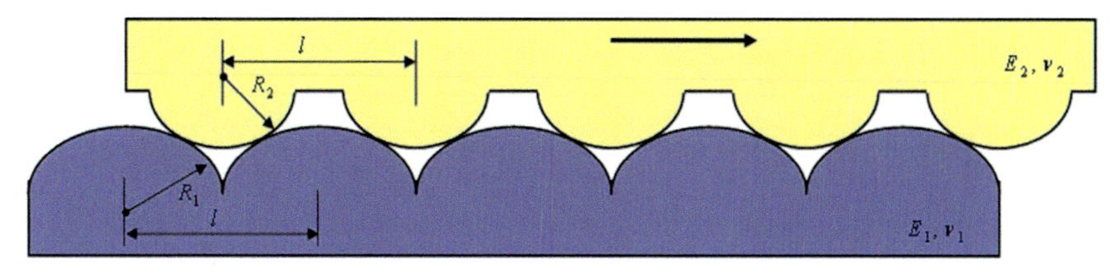

Figure 4.13 Scheme of mutual sliding of two regular surfaces.

4.2.3 Example of the friction force calculation

The results of calculation of the adhesion friction force (dimensional tangential stress) are presented in Fig. 4.14 for sliding of two elastic bodies with the reduced modulus of elasticity $E^* = 10$ MPa and adhesion pressure $p_0 = 1$ MPa, which correspond to some elastomeric materials.

In Fig. 4.14a, the results are presented for the reduced radius of asperities $R = 0.1$ mm and the period length $l = 0.1$ mm. The value of the specific work of adhesion w_a ranges from 0 to 0.2 J/m^2, which corresponds to the adhesion parameter λ ranging from infinity to about 4.

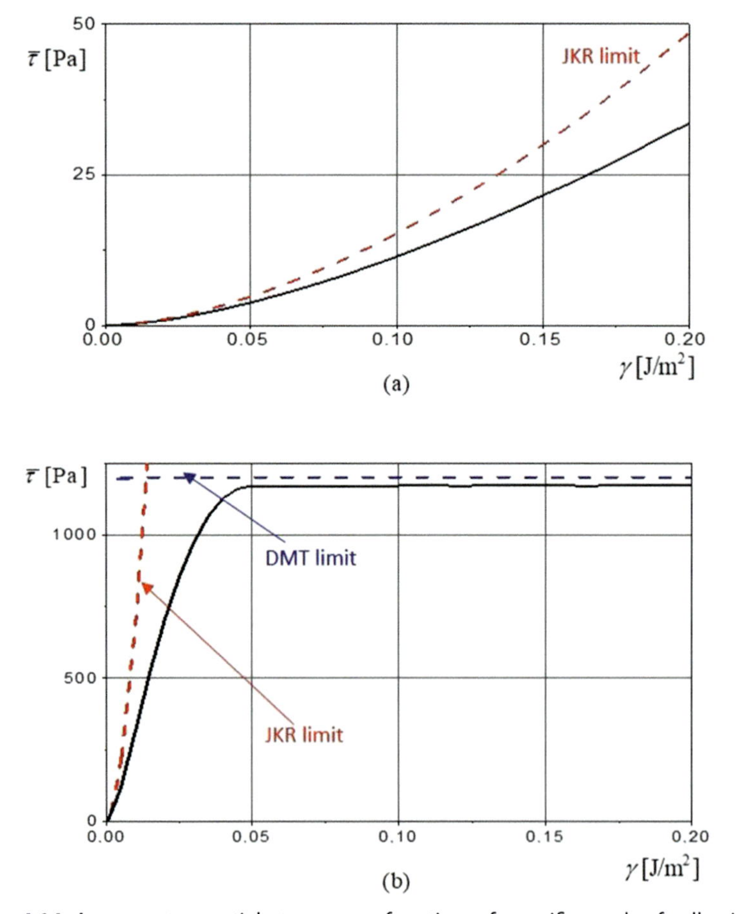

Figure 4.14 Average tangential stress as a function of specific work of adhesion in sliding of two surfaces with regular relief for the cases of larger asperities (a) and smaller asperities (b).

The results presented in Fig. 4.14b correspond to surfaces with finer relief, $R = l = 1$ μm. In this case, the specific work of adhesion w_a ranging from 0 to 0.2 J/m^2 corresponds to the adhesion parameter λ ranging from infinity to approximately 0.4.

Dashed lines in Fig. 4.14 correspond to the limit cases, in which analytic relations for the friction force are obtained. As $\lambda \to \infty$ (the JKR limit), the dimensionless energy loss is defined by Eq. (4.14). In accordance with Eq. (4.31) dimensionless tangential stress is calculated as

$$\overline{\tau}_{JKR} = \frac{3^{1/3}\left(2^{1/3}+12\right)}{15l^3}\left(\frac{9\pi^5 w_a^5 R^4}{16E^{*2}}\right)^{1/3} \approx \frac{7.09}{l^3}\left(\frac{w_a^5 R^4}{E^{*2}}\right)^{1/3}. \tag{4.33}$$

For small values of the parameter λ ($\lambda < (9/32)^{1/3} \approx 0.66$), which can be considered the DMT limit, the energy loss is determined by Eq. (4.11). Taking into account Eq. (4.31), we obtain

$$\overline{\tau}_{DMT} = 12\frac{p_0^5 R^3}{l^3 E^{*4}}. \tag{4.34}$$

The results indicate that in the entire range of values of the adhesion parameter, including the limit cases, the adhesion component of the friction force is higher for asperities with larger radius of curvature R and for softer materials.

4.3 Modeling of rolling friction force

The energy dissipation in formation and breaking of elementary contacts is also a mechanism of the molecular component to the rolling friction resistance. Experimental and numerical evaluation of this component of the rolling friction force was first given (Tomlinson, 1929) based on the assumption that each approach and retraction of molecules is associated with an energy loss. Derjaguin and Toporov (1994) explained the rolling resistance by attraction of the surface parts moving away from each other, which was accounted for by electrical charges arising on the surfaces in contact.

In this section, the adhesive component of the rolling friction force is determined based on the energy dissipation mechanism occurring when asperities of the rolling body approach the foundation and then retract from it (Goryacheva and Makhovskaya, 2007, 2017).

4.3.1 Problem formulation for a rough cylinder

Consider a rough cylinder of the radius R (1) rolling with the angular velocity ω on an elastic half-space (2); see Fig. 4.15. The problem is considered in the moving system of coordinates, the z-axis of which is directed downward inside the elastic half-space, and the x-axis coincides with the undeformed surface of the elastic half-space and is directed in the direction of motion of the cylinder. The surface of the cylinder is covered with a periodic system of asperities located in nodes of the quadratic lattice with a pitch l. The tops of all asperities have the same shape with the radius of curvature R_0, where $R_0 << R$. The normal external force P is applied at the axis of the cylinder.

The gap between the surfaces of the rough cylinder and the elastic half-space is given by the relation

$$h(x, y) = u_z(x, y) + f(x, y) - c, \tag{4.35}$$

where $u_z(x, y)$ is the elastic displacement of the surface of the half-space in the z-direction, $f(x, y)$ is a function defining the shape of the rough cylinder surface, and c is the maximum penetration of the cylinder into the elastic half-space.

The contact between the cylinder and the half-space occurs over the regions A_i, in which the contact condition is satisfied:

$$h(x, y) = 0, \ (x, y) \in A_i. \tag{4.36}$$

Tangential stresses on the surface of the half-space are assumed to be zero.

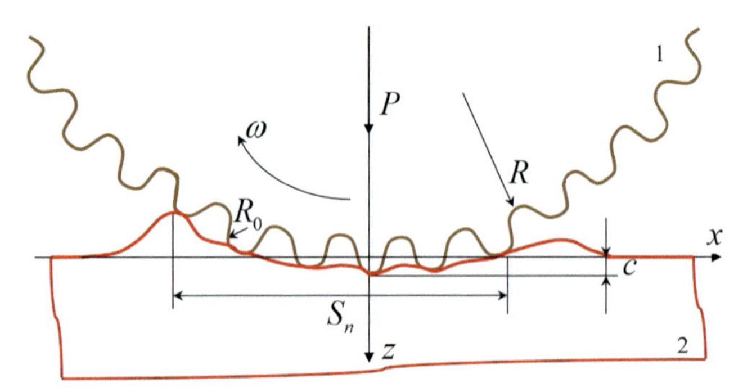

Figure 4.15 Scheme of rolling of a rigid rough cylinder (1) over the elastic half-space (2).

The surfaces of the cylinder and half-space are attracted to each other by the force of molecular adhesion. This attraction occurs in the regions B_i which are either ring-shaped (for asperities contacting with the half-space) or circular (for asperities that are not in direct contact with the half-space). The specific force of adhesion (adhesive pressure) acting between the surfaces is described by the relation that follows from Eq. (2.6):

$$p_a(x,y) = \begin{cases} -p_0, & h(x,y) \le h_0 \\ 0, & h(x,y) > h_0 \end{cases}, \quad (x,y) \in B_i. \tag{4.37}$$

The specific work of adhesion is defined by Eq. (2.7).

4.3.2 Reducing to the problem for a separate asperity

To solve the problem, we assume that the stress-strain state of the half-space near each asperity is independent of the influence of other asperities. This assumption is valid if asperities are sufficiently distant from each other. In this case, the interaction of each asperity with the half-space can be considered separately.

We consider the configuration in which the lowest asperity of the rough cylinder is located symmetrically with respect to the z-axis (see Fig. 4.15). Let the indentation of this asperity into the elastic half-space, which coincides with the maximum indentation of the cylinder, be c, and let the indentation of an i-th asperity be c_i (Fig. 4.16), where i is an integer ranging from 1 to N_{app} to the right of the central asperity (N_{app} is the number of asperities approaching the half-space) and from 1 to N_{sep} to its left (N_{sep} is the number of asperities moving away from the half-space). In total, $N_{\text{app}} + N_{\text{sep}} + 1$ asperities interact with the elastic half-space.

From the triangles ABC and AOC (O is the center of the cross-section of the cylinder, A is the top point of an i-th asperity, C is the top point of the central asperity), we have the following relation for the length of the AC segment:

$$AC = \frac{c - c_i}{\sin(\psi_i/2)} = 2R\sin(\psi_i/2). \tag{4.38}$$

By assuming that $l << R$, we obtain the angle $\psi_i = il/R$. Then the indentation of an i-th asperity into the half-space is given by

$$c_i = c - 2R\sin^2\frac{il}{2R}. \tag{4.39}$$

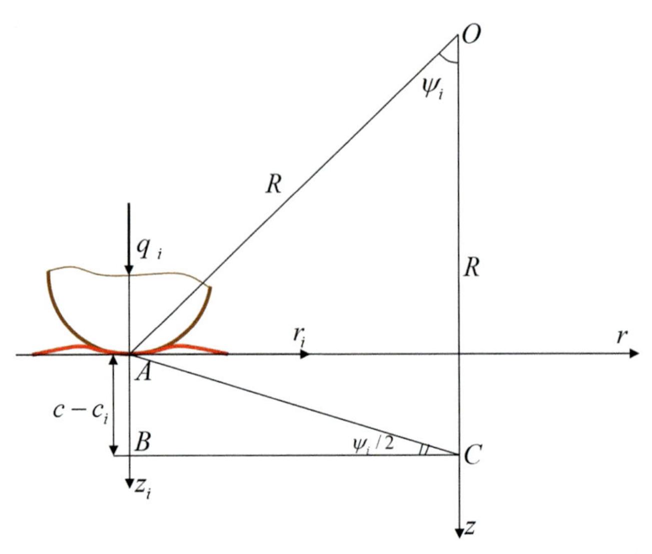

Figure 4.16 Scheme of contact of a separate asperity of the rough cylinder with the elastic half-space.

Since the indentation c_i is known for each asperity, we can consider the interaction of an i-th asperity with the half-space separately and independently of other asperities. Let us introduce a local cylindrical system of coordinates (r_i, ϕ_i, z_i) with the origin at the center of the i-th contact region. Let the asperity have a semispherical shape; then the distributions of stresses and displacement in the half-space near this asperity are symmetrical with respect to the z_i-axis (Fig. 4.16). Inside the contact region A_i, which is a circle of a radius a_i, the contact condition is satisfied for the boundary displacement u_z along the axis z_i. In virtue of Eq. (4.36), this condition has the form:

$$u_z(r_i) = -\frac{r_i^2}{2\rho} + c_i, \quad r_i \le a_i. \tag{4.40}$$

From condition Eq. (4.37) and symmetry of the problem, it follows that inside the region of adhesion B_i, which is the ring $a_i < r_i < b_i$, the pressure P on the surface of the elastic half-space is defined by the adhesion attraction

$$p(r_i) = \begin{cases} -p_0, & a_i < r_i < b_i \\ 0, & r_i > b_i \end{cases}. \tag{4.41}$$

Normal pressure $p(r_i)$ and displacement $u_z(r_i)$ are related to each other by Eq. (2.19).

Eq. (2.19) with conditions Eqs. (4.40) and (4.41) allows us to determine the normal contact pressure $p(r_i)$ and displacement $u_z(r_i)$ near the contact of an i-th asperity and the half-space. To calculate the unknown values a_i and b_i, it is necessary to use the following condition of continuity of the pressure at the contact region boundary,

$$p(a_i) = -p_0, \tag{4.42}$$

and the condition following from Eq. (4.41) and the continuity of the local gap h_i between the half-space and the i-th asperity:

$$p_0 h(b_i) = w_a, \quad \text{where} \quad h(b_i) = u_z(r_i) + \frac{r_i^2}{2R_0} - c_i. \tag{4.43}$$

Eqs. (2.19) and (4.40)−(4.43) allow us to determine the contact pressure and elastic displacement of the half-space near an i-th asperity, whose indentation is given by Eq. (4.39). After this, the normal force acting on each asperity is calculated as

$$q_i = 2\pi \int_0^{b_i} rp(r)dr. \tag{4.44}$$

4.3.3 Solution for a separate asperity

The solution of the problem of adhesive interaction between two axisymmetric elastic bodies was given in Section 4.1. For the case where one of these bodies is a spherical asperity of the radius R and the other is the elastic half-space ($n = 1$, $A = 1/(2R)$), from Eq. (4.2) we obtain the equation relating the radii of the regions of contact and adhesion, a_i and b_i:

$$\frac{4b_i p_0^2}{\pi E^*}\left(1 - \alpha_i + \phi_i\sqrt{1-\alpha_i^2}\right) - \frac{b_i^2 p_0}{\pi R_0}\left[(2\alpha_i^2 - 1)\phi_i + \alpha_i\sqrt{1-\alpha_i^2}\right] + w_a = 0, \tag{4.45}$$

where the following notation is used: $\alpha_i = a_i/b_i$, $\phi_i = \arccos \alpha_i$.

From Eq. (4.1) we obtain the expressions for the indentation of an asperity and the force acting upon it:

$$c_i = -2b_i\left(\frac{p_0}{E^*}\sqrt{1-\alpha_i^2} + \alpha_i^2\right), q_i$$

$$= \frac{4E^*}{3R_0}b_i^3\alpha_i^3 - p_0 b_i^2\left(2\alpha_i\sqrt{1-\alpha_i^2} - \frac{\phi_i}{2}\right). \tag{4.46}$$

By solving Eq. (4.45) for b_i and substituting the result into Eq. (4.46), we obtain the parametric dependence between the normal force q_i acting on an asperity and the indentation c_i of this asperity, α_i being the parameter.

In the case of contactless interaction between an asperity and the half-space, from Eq. (4.4) we obtain the dependence between q_i and c_i in the form

$$c_i = \frac{b_i^2}{2R_0} - \frac{4p_0 b_i}{\pi E^*} - \frac{w_a}{p_0}, \quad q_i = -\pi b_i^2 p_0. \tag{4.47}$$

The normal force q_i acting on an asperity as a function of its indentation c_i is illustrated in Fig. 4.17. The thick line corresponds to the contact of the surfaces and Eqs. (4.45) and (4.46). The thin line corresponds to the contactless interaction and Eq. (4.47). As the indentation c_i is increased from $-\infty$ (the asperity approaches the half-space) the surfaces jump into contact at $c_i = c^{\text{app}}$, after which the interaction goes along path 1. Let us denote the corresponding force-indentation dependence by $q_i^{\text{app}}(c_i)$. As the indentation is decreased (the asperity is retracted from the half-space), the interaction of surfaces breaks at $c_i = c^{\text{sep}}$ (path 2). Let the corresponding force-indentation dependence be denoted by $q_i^{\text{sep}}(c_i)$. The functions $q_i^{\text{app}}(c_i)$ and $q_i^{\text{sep}}(c_i)$ are determined by Eqs. (4.45)–(4.47) in which the corresponding branch of the ambiguous solution should be chosen.

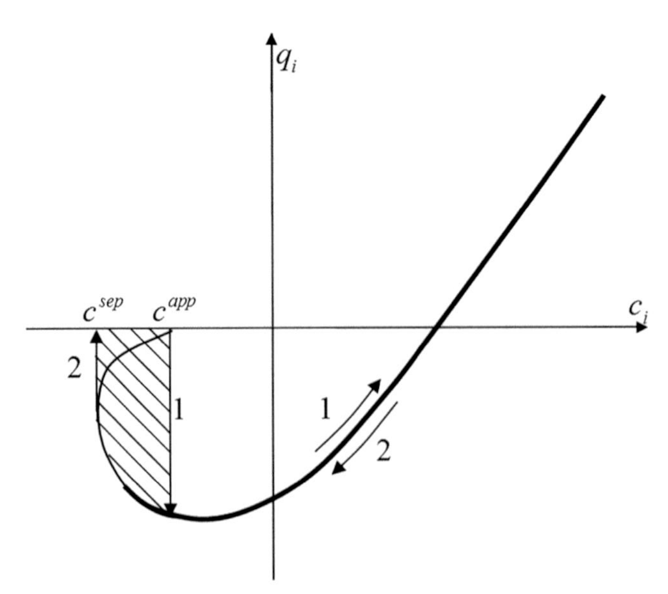

Figure 4.17 Normal force acting on an asperity of the rough cylinder as a function of its indentation into the elastic half-space.

4.3.4 Solution for a rough cylinder

The obtained force-indentation dependencies for an asperity can be applied to construct the solution for a rolling rough cylinder. For a prescribed maximum indentation c of the cylinder into the half-space, which coincides with the indentation of the central asperity, the indentations c_i of the remaining asperities are defined by Eq. (4.39). By using the solution for one asperity, we can thus calculate the corresponding normal forces q_i acting on each asperity from the half-space (as it follows from Fig. 4.17, these forces can be positive or negative) and radii of the regions of contact a_i and adhesion b_i. The force acting on an i-th asperity depends on the value c_i and on whether this asperity approaches the half-space or moves away from it. The number of asperities interacting with the half-space to the right of the central asperity (asperities approaching the half-space in rolling) is defined by the geometrical equation

$$c - c^{\mathrm{app}} = 2R\,\sin^2\frac{N_{\mathrm{app}}l}{2R}, \tag{4.48}$$

from which we have

$$N_{\mathrm{app}} = \frac{l}{2R}\arcsin\sqrt{\frac{c - c^{\mathrm{app}}}{2R}}. \tag{4.49}$$

In a similar way, we determine the number of asperities interacting with the half-space to the left of the central asperity (retracting from the half-space in rolling):

$$N_{\mathrm{sep}} = \frac{l}{2R}\arcsin\sqrt{\frac{c - c^{\mathrm{sep}}}{2R}}. \tag{4.50}$$

In calculating by Eqs. (4.49) and (4.50), one should take the maximum integer not exceeding the value of the right-hand side. The condition $N_{\mathrm{app}} \le N_{\mathrm{sep}}$ is always satisfied.

Thus, if the indentation c of the central asperity is known, one can calculate indentations of all asperities by Eq. (4.39) and after this from Eqs. (4.45)–(4.47) determine the relationship between the force q_i for each asperity and the indentation c_i. From this relationship, one calculates the quantities c^{app} and c^{sep}, as well as the functions $q_i^{\mathrm{app}}(c_i)$ and $q_i^{\mathrm{sep}}(c_i)$. In a similar way, one determines the contact radii for each asperity as functions of the indentation: $a_i^{\mathrm{app}}(c_i)$ and $a_i^{\mathrm{sep}}(c_i)$. The functions obtained make it possible to calculate the values of the forces acting on each asperity, q_i^{app}

$(i = 1..N_{app})$ on the right of the central asperity and q_i^{sep} $(i = 1...N_{sep})$ on the left of the central asperity. The force acting on the central asperity can be calculated from any of the functions obtained: $q_0 = q_0^{app}(c) = q_0^{sep}(c)$. Then the total normal force acting on the cylinder is calculated as

$$P = q_0 + \sum_{i=1}^{N_{app}} q_i^{app} + \sum_{i=1}^{N_{sep}} q_i^{sep}. \tag{4.51}$$

4.3.5 Analysis of the contact characteristics

The results of the calculation of the forces acting on asperities of the rough cylinder are shown in Fig. 4.18 by curve 1, which represents the distribution of the dimensionless nominal pressure $q_i/(lE^*)$ over the dimensionless coordinate $x_i/l = li/R$. The results are calculated for the dimensionless specific work of adhesion $w_a/(p_0R_0) = 0.01$, the adhesive pressure $p_0/E^* = 0.1$, and the normal external force applied to the cylinder $P/(E^*R^2) = 1.25 \times 10^{-4}$. The ratio of the asperities radius to cylinder radius is $R_0/R = 0.01$, the number of asperities in the cross-section of the cylinder is $N = 5000$.

The calculations show that in this case, the number of asperities interacting with the half-space is $N_{app} = 46$ on the right and $N_{sep} = 53$ on the left. The graph presented in Fig. 4.18 (curve 1) shows that the nominal pressure is positive in the center of the nominal contact area and it becomes

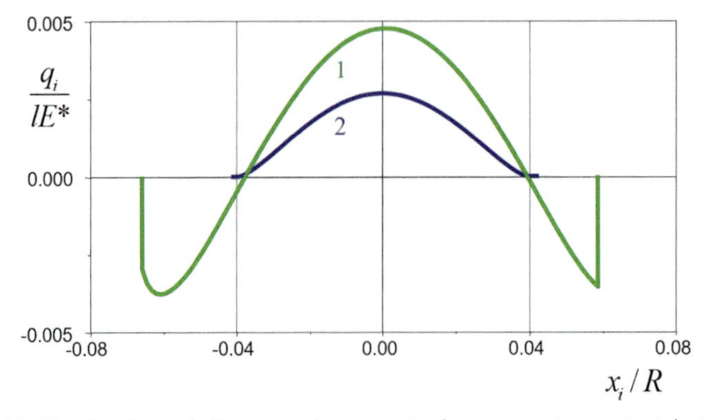

Figure 4.18 Distribution of dimensionless nominal pressure in contact between the rough rolling cylinder and the half-space.

negative near its boundary. The distribution is nonsymmetric, which leads to the moment of resistance to rolling. For comparison, the distribution of nominal pressure is presented for the case of no adhesion (curve 2); in this case the pressure is positive over the entire nominal contact region, its distribution is symmetric, and the moment of rolling resistance is zero.

The solution obtained also makes it possible to calculate the real contact area per unit length of the cylinder:

$$S_r = \pi\left(a_0^2 + \sum_{i=1}^{N_{\text{app}}} (a_i^{\text{app}})^2 + \sum_{i=1}^{N_{\text{sep}}} (a_i^{\text{sep}})^2\right). \tag{4.52}$$

And the corresponding nominal contact area is as follows:

$$S_n = l^2\left(1 + N_{\text{app}} + N_{\text{sep}}\right). \tag{4.53}$$

In Fig. 4.19, the dimensionless nominal contact area S_n/R^2 (dashed lines) and real contact area S_r/R^2 (solid lines) are presented as functions of the normal force applied to the cylinder $P/\left(E^* R^2\right)$ in the presence of adhesion specified by the parameters $w_a/p_0R_0 = 0.01$, $p_0/E^* = 0.1$ (curves 1) and in the absence of adhesion (curves 2). Geometrical characteristics of the rough cylinder are $R_0/R = 0.01$, $N = 10^4$. The results indicate that taking into account adhesion leads to ambiguous dependence of both real and nominal contact areas on the load, and also to the existence of contact area under negative loads.

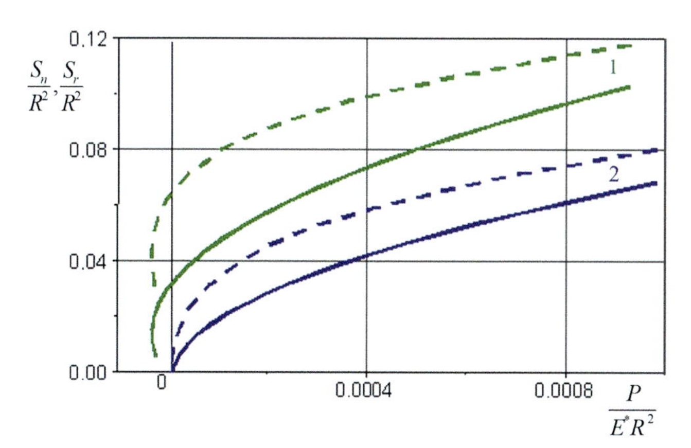

Figure 4.19 Nominal (*dashed lines*) and real (*solid lines*) contact areas versus external normal force acting on the rough rolling cylinder in the presence of adhesion (*lines 1*) and without adhesion (*lines 2*).

4.3.6 Calculation of rolling resistance

Since the dependence of the force q_i on the indentation c_i of an asperity is ambiguous (Fig. 4.17), an energy loss occurs when this asperity cyclically moves to the elastic half-space and apart from it, provided that the maximum indentation of the asperity into the half-space exceeds c^{app}. The value of this energy loss corresponds to the area of the dashed region in Fig. 4.17 and is calculated as

$$\Delta w = \int_{c^{\mathrm{app}}}^{c^{\mathrm{sep}}} \left[q_i^{\mathrm{app}}(c) - q_i^{\mathrm{sep}}(c)\right] dc. \tag{4.54}$$

The energy lost in a revolution of the cylinder is $\Delta w N_1$, where N_1 is the number of asperities in the cylinder cross-section, for which the maximum indentation exceeds the value c^{app}. By assuming that this energy loss is equal to the work of the moment of rolling resistance $2\pi M$ per one revolution of the cylinder, we obtain the following relation for the moment of rolling resistance:

$$M = \frac{\Delta w N_1}{2\pi}. \tag{4.55}$$

If the cylinder has N asperities in the cross-section, then the number N_1 is given by the stepwise function:

$$N_1 = \begin{cases} N, & c \geq c^{\mathrm{app}} \\ 0, & c < c^{\mathrm{app}} \end{cases}. \tag{4.56}$$

If the asperities have statistical distribution of heights, then we have

$$N_1 = N \int_{-\infty}^{c} \phi(t)\, dt, \tag{4.57}$$

where $\phi(t)$ is the density of distribution, e.g., according to the Gauss law:

$$\phi(t) = \frac{1}{\sqrt{2\pi}\sigma} \exp\left(-\frac{t^2}{2\sigma^2}\right). \tag{4.58}$$

Plots of the dimensionless moment $M/(E^*R^3)$ of rolling resistance versus the distance $d = -c$ between the cylinder and the half-space are presented in Fig. 4.20 for asperities of the same height (curve 1) and for the

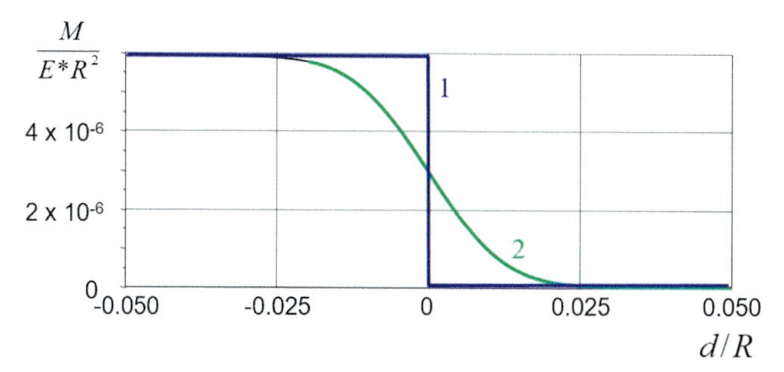

Figure 4.20 Dimensionless moment of rolling resistance as a function of distance between the cylinder and the half-space for the same height asperities (*curve 1*) and Gauss distribution of heights (*curve 2*).

Gauss distribution of their heights (curve 2). The results are calculated for $w_a/(p_0R_0) = 0.1$, $p_0/E^* = 0.1$, $R_0/R = 0.01$, $N = 10^4$, and the mean square deviation for the case of the Gauss distribution of heights is $\sigma/R = 0.01$. The results indicate that as the indentation of the cylinder into the half-space increases, the increase in the rolling resistance is stepwise in the first case and smooth in the second case. For both cases, the moment of rolling resistance tends to the same constant values for large indentations. This value depends on the geometric parameters of the cylinder, elastic properties of the half-space, and adhesive properties of their surfaces.

The results obtained also show that the moment M of rolling resistance increases with increasing specific work of adhesion w_a. This is illustrated by graphs presented in Fig. 4.21, which are obtained for $p_0/E^* = 0.1$,

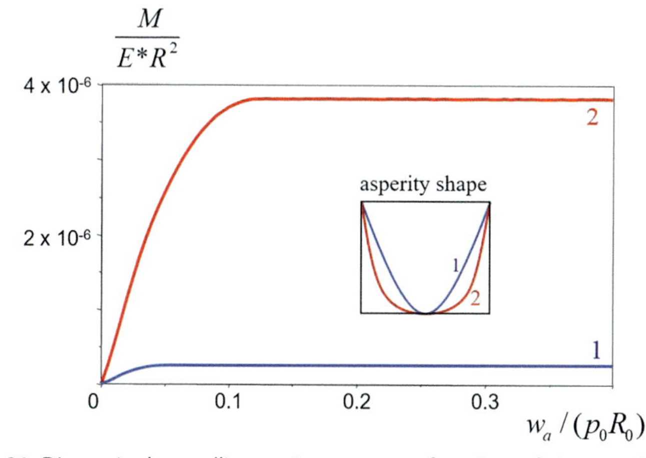

Figure 4.21 Dimensionless rolling resistance as a function of the specific work of adhesion for two different shapes of asperities.

$N = 10^4$ and two different shapes of asperities, $n = 1$ (curve 1) and $n = 2$ (curve 2) for the case of the same heights of asperities. The moment of rolling resistance tends to a constant value as the specific work of adhesion increases. The rolling resistance is higher for asperities with flatter top.

4.4 Conclusion

The energy loss in an approach-retraction cycle of two axisymmetric elastic asperities is calculated for the cases of capillary and molecular adhesion. The value of this energy loss is analyzed depending on the elastic properties of the asperities, volume of fluid in the meniscus, and specific work of adhesion of dry surfaces. It is established that for the asperities having the shape of paraboloids of revolution, the dimensionless energy loss is the function of only one parameter. This parameter is established both for capillary and for molecular adhesion cases. In the case of molecular adhesion, the dimensionless energy loss is calculated in the closed form in the limit cases of small λ (DMT approximation) and large λ (JKR approximation).

A method is developed to calculate the adhesive component of the friction force in the conditions of sliding and rolling of elastic bodies with rough surfaces. The case of molecular adhesion of dry surfaces is considered. The method is based on the determination of the energy dissipation in an approach-retraction cycle of asperities of rough surfaces. The friction force is calculated for various parameters of surface relief and specific work of adhesion of two interacting surfaces. The results obtained can be applied for governing the parameters of microgeometry of surfaces to attain required frictional characteristics for sliding and rolling contact.

References

Barquins, M., 1990. Adherence, friction and contact geometry of a rigid cylinder rolling on the flat and smooth surface of an elastic body. J. Nat. Rubber Res. 5 (3), 199–210.

Carbone, G., Mangialardi, L., 2004. Adhesion and friction of an elastic half-space in contact with a slightly wavy rigid surface. J. Mech. Phys. Solid. 52 (6), 1267–1287. https://doi.org/10.1016/j.jmps.2003.12.001.

Chaudhury, M.K., Owen, M.J., 1993. Adhesion hysteresis and friction. Langmuir 9 (1), 29–31. https://doi.org/10.1021/la00025a009.

Chekina, O.G., 1998. On friction of rough surfaces separated by a thin liquid layer. J. Frict. Wear 19 (3), 306–311.

Chizhik, S.A., 1994. Capillary mechanism of adhesion and friction of rough surfaces separated by a thin layer of liquid. J. Frict. Wear 15 (1), 11–26.

Derjaguin, B.V., Toporov, Y.P., 1994. Influence of adhesion on the sliding and rolling friction. Prog. Surf. Sci. 45 (1–4), 317–327. https://doi.org/10.1016/0079-6816(94)90064-7.

Goryacheva, I.G., Gubenko, M.M., Makhovskaya, Y.Y., 2014. Sliding of a spherical indenter on a viscoelastic foundation with the forces of molecular attraction taken into account. J. Appl. Mech. Tech. Phys. 55 (1), 81–88. https://doi.org/10.1134/S0021894414010118.

Goryacheva, I.G., Makhovskaya, Y.Y., 2001. Adhesive interaction of elastic bodies. J. Appl. Math. Mech. 65 (2), 273–282. https://doi.org/10.1016/S0021-8928(01)00031-4.

Goryacheva, I.G., Makhovskaya, Y.Y., 2007. Adhesive resistance in the rolling of elastic bodies. J. Appl. Math. Mech. 71 (4), 485–493. https://doi.org/10.1016/j.jappmathmech.2007.09.001.

Goryacheva, I., Makhovskaya, Y., 2011. A model of the adhesive component of the sliding friction force. Wear 270 (9–10), 628–633. https://doi.org/10.1016/j.wear.2011.01.020.

Goryacheva, I., Makhovskaya, Y., 2017. Combined effect of surface microgeometry and adhesion in normal and sliding contacts of elastic bodies. Friction 5 (3), 339–350. https://doi.org/10.1007/s40544-017-0179-1.

Greenwood, J.A., 1997. Adhesion of elastic spheres. Proc. Math. Phys. Eng. Sci. 453 (1961), 1277–1297. https://doi.org/10.1098/rspa.1997.0070.

Grigoriev, A.M., Dubravin, A.V., Kovalev, I.N., Kovaleva, O.Y., Komkov, O.Y., Myshkin, N.K., 2003. Measurement of contact adhesion and attraction between the engineering surfaces. J. Frict. Wear 24, 51–58.

Heise, R., Popov, V.L., 2010. Adhesive contribution to the coefficient of friction between rough surfaces. Tribol. Lett. 39 (3), 247–250. https://doi.org/10.1007/s11249-010-9617-1.

Jacquot, C., Takadoum, J., 2001. A study of adhesion forces by atomic force microscopy. J. Adhes. Sci. Technol. 15 (6), 681–687. https://doi.org/10.1163/156856101750430422.

Johnson, K.L., 1997. Adhesion and friction between a smooth elastic spherical asperity and a plane surface. Proc. R. Soc. London. Ser. A. 453, 163–179. https://doi.org/10.1098/rspa.1997.0010.

Maugis, D., 1992. Adhesion of spheres: the JKR-DMT transition using a Dugdale model. J. Colloid Interface Sci. 150, 243–269. https://doi.org/10.1016/0021-9797(92)90285-T.

Szoszkiewicz, R., Bhushan, B., Huey, B., Kulik, A., Gremaud, G., 2005. Correlations between adhesion hysteresis and friction at molecular scales. J. Chem. Phys. 122 (14), 144708. https://doi.org/10.1063/1.1886751.

Tomlinson, J.A., 1929. A molecular theory of friction. Philos. Malacit. Supl. 7 (46), 905–939.

Wei, Z., Zhao, Y.P., 2004. Adhesion elastic contact and hysteresis effect. Chin. Phys. 13 (8), 1320–1325. https://doi.org/10.1088/1009-1963/13/8/024.

Yoshizawa, H., Chen, Y.L., Israelachvili, J., 1993. Fundamental mechanisms of interfacial friction. 1. Relation between adhesion and friction. J. Phys. Chem. 97 (16), 4128–4140. https://doi.org/10.1021/j100118a033.

CHAPTER 5

Microgeometry effect in sliding contact of viscoelastic solids

In this chapter, solutions of contact problems for viscoelastic bodies in sliding conditions are presented taking into account their surface microgeometry. Effect of a thin surface viscoelastic layer bonded to the elastic half-plane on the contact characteristics and friction force is analyzed in 2-D periodic problem formulation. An analytic solution of the problem for a wavy indenter sliding over the viscoelastic half-plane is presented for saturated contact mode. By using the strip method, 3-D periodic contact problems for a viscoelastic foundation are analyzed, including problems with adhesion and incompressible fluid in the gap. To take into account multiscale features of the surface microgeometry, a model of the multiscale surface waviness is developed, and the local friction law is taken into account, which is a result of hysteretic loss combined with adhesion in sliding contact at inferior scale levels.

5.1 Hysteretic mechanism of energy dissipation

In friction of viscoelastic materials, such as rubbers, highly elastic polymers, and composites based on these polymers, one of the major mechanisms of energy dissipation is viscoelastic hysteresis occurring in cyclic deformation of surface layers of materials (Greenwood and Tabor, 1958; Grosch, 1963). To describe this process, models of viscoelastic bodies are used. An important factor that significantly affects the hysteretic mechanism is microgeometry of contacting surfaces, which determines the frequency of interaction of the material and duration of its contact with a counterbody (Fig. 5.1a). Analysis of the combined action of these factors is important when developing methods of control of the friction force in contact of bodies with imperfect elasticity of surface layers.

To study these mechanisms, contact problems in quasistatic formulation were solved for a viscoelastic material in sliding contact with a counterbody whose surface microrelief was described in various ways. 2-D contact

Discrete Contact Mechanics with Applications in Tribology
ISBN 978-0-12-821799-3
https://doi.org/10.1016/B978-0-12-821799-3.00005-4

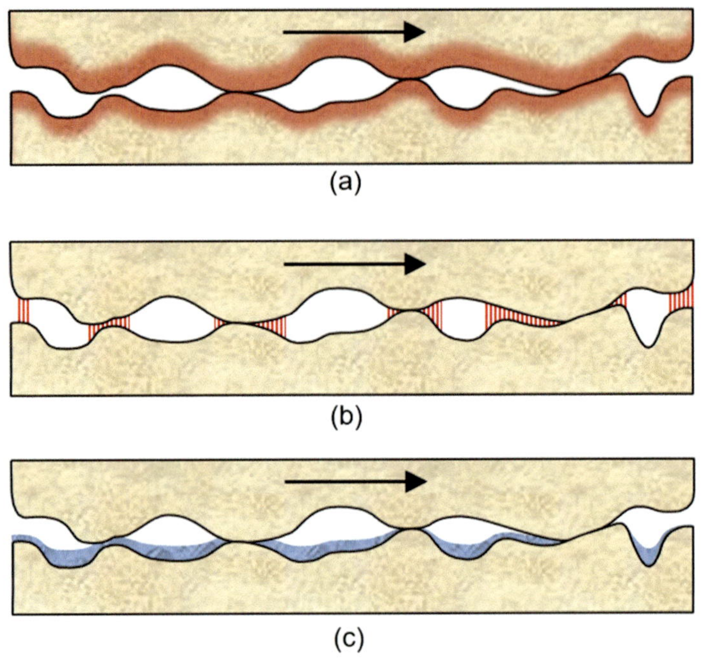

Figure 5.1 Illustration of sliding of two solids with microgeometry in the presence of hysteresis in surface layers (a), adhesion (b), fluid in the gap (c).

problems were considered for a periodic system of punches moving over a thin viscoelastic layer bonded to an elastic half-space. Deformation of the layer was described by the Maxwell model (Goryacheva and Sadeghi, 1995; Goryacheva et al., 1996) and by the Kelvin model (Goryacheva and Makhovskaya, 1997). An analytic solution was presented for the contact of a rigid sinusoidal punch and a viscoelastic half-plane (Menga et al., 2014), which was based on the Hunter method for a contact region (Hunter, 1961). The results showed that depending on the nominal pressure and sliding velocity, transition from partial to saturated contact occurred. The hysteretic component of the coefficient of friction, size of the contact region, and its shift were obtained as functions of rheologic and elastic properties of the half-plane material, geometric parameters of the wavy surface, nominal pressure, and sliding velocity.

3-D contact problems were also considered and solved in quasistatic formulation. The contact problem for a rigid cylinder sliding over a wavy surface of the viscoelastic layer described by the Kelvin model was considered under the assumption that the cylinder axis was perpendicular to the surface waviness cross-section (Goryacheva and Goryachev, 2020).

A 3-D contact problem was studied for a doubly periodic wavy surface sliding over a viscoelastic layer described by the Kelvin model with a spectrum of relaxation times (Sheptunov et al., 2013).

Another important mechanism contributing into the friction of viscoelastic bodies is the adhesion force acting in the gap around real contact spots (Fig. 5.1b). This contribution is particularly significant in interaction of dry and clean surfaces at micro- and nano-scale levels. The method of solution of contact problems for a viscoelastic foundation in sliding contact with a rigid wavy surface was suggested for 2-D case (Goryacheva and Makhovskaya, 2010) and for 3-D case (Goryacheva and Makhovskaya, 2015, 2016). The formulation of those problems implied only normal forces acting in the contact, the tangential force (friction force) being calculated as a result of hysteretic response. It was assumed that the adhesion forces acting in normal direction to the surfaces cause no adhesive friction component, but they only influence the hysteretic friction through the contact stress redistribution. This model was able to explain the difference in experimentally measured friction coefficients for rubber samples with different adhesive properties (Morozov and Makhovskaya, 2019). Sliding of a multilevel 3-D wavy surface over a viscoelastic foundation was considered (Makhovskaya, 2019). At each scale level, the friction force was calculated as a result of hysteretic losses at this level with the additional term defined by the local friction law obtained at the inferior scale level. In this chapter, we use these approaches to analyze the combined influence of adhesion, surface microgeometry, and viscoelastic hysteresis on the contact characteristics and sliding friction force.

When a rough surface slides over a viscoelastic body in the presence of fluid in the gap, e.g., a car tire tread slips over a wet road coating, the contribution of the adhesion mechanism into the total friction force is insignificant. In this case, the friction force is influenced by the fluid behavior (Fig. 5.1c). Depending on the amount of fluid and sliding velocity, the mechanism of this influence can be different, from capillary effects in a very thin fluid layer to hydrodynamic effects when a fluid layer completely separates the surfaces. In this chapter, an intermediate situation is considered in which fluid fills the gap between the surfaces, but it does not penetrate into contact regions; i.e., it does not completely separate the surfaces. In the conditions of normal loading of a surface with regular relief in contact with an elastic half-space, the effect of fluid in the gap on the contact characteristics was studied previously (Kuznetsov, 1985). Below, based on the solution of the problem for a rigid 3-D periodical punch sliding over a viscoelastic foundation in the presence of incompressible fluid in the gap (Goryacheva and Shpenev, 2012), we will analyze the effect of fluid on the resistance to the punch motion and contact pressure distribution.

5.2 Effect of microgeometry and thin viscoelastic layer in sliding contact of elastic solids (2-D analysis)

In contact interaction, particular properties of surface layers considerably influence the contact characteristics, friction, and wear. In this section, the combined effect of a surface microgeometry and an imperfect elasticity of a surface layer in sliding contact is studied.

5.2.1 Problem formulation

A 2-D contact problem is considered for a rigid indenter with periodically arranged asperities that slides with a constant velocity V over the surface of a thin viscoelastic layer of thickness H bonded to the elastic base (Fig. 5.2).

The indenter shape is described by a periodic (with a period l) function $f(x)$, i.e., $f(x) = f(x + l)$. For the viscoelastic layer, two models were

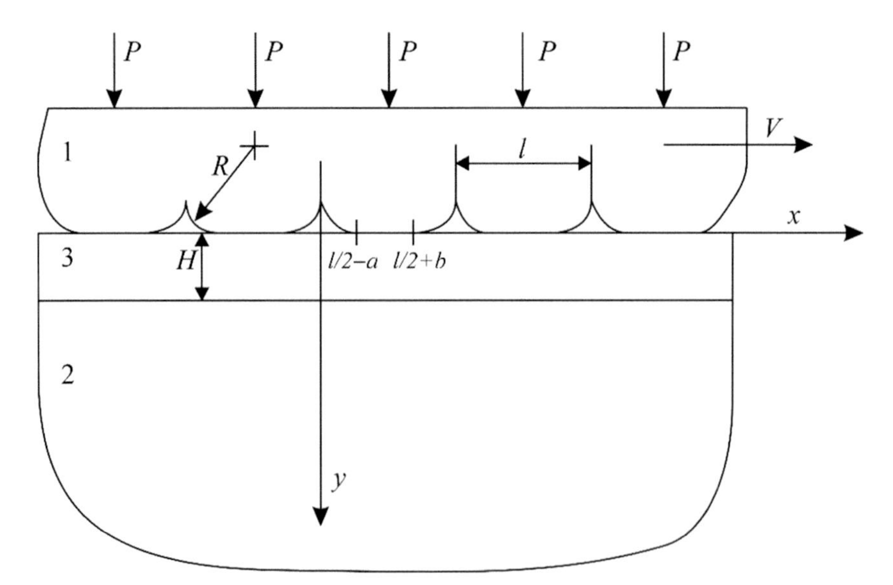

Figure 5.2 Scheme of sliding contact between the periodic indenter (1) and viscoelastic layer (3) bonded to the elastic half-plane (2).

previously considered: the Maxwell model (Goryacheva and Sadeghi, 1995; Goryacheva et al., 1996) and the Kelvin model (Goryacheva and Makhovskaya, 1997). For the Kelvin model, which is characterized by limited creep, the relation between the displacement $u_1(x,t)$ of the upper boundary of the layer in the y-direction and the normal pressure $p(x,t)$ has the following form:

$$u_1 + T_\varepsilon \frac{\partial u_1}{\partial t} = \frac{H}{E_1}\left(p + T_\sigma \frac{\partial p}{\partial t}\right), \tag{5.1}$$

where T_ε and T_σ are the retardation and relaxation times, E_1 is the long-term modulus of elasticity, and H is the thickness of the layer.

The motion is assumed to be steady state. Let the moving system of coordinates (x,y) be attached to the indenter. In the system of coordinates (x,y), the normal displacement $u_1(x)$ and pressure $p(x)$ are independent of time and are functions of only the coordinate x. In the moving coordinate system, Eq. (5.1) has the form

$$u_1 - T_\varepsilon V \frac{du_1}{dx} = \frac{H}{E_1}\left(p - T_\sigma V \frac{dp}{dx}\right). \tag{5.2}$$

The normal displacement of the layer satisfies the condition of periodicity:

$$u_1(x) = u_1(x+l). \tag{5.3}$$

Since for the 1-D model of the viscoelastic layer, pressure does not change over the thickness of the layer, the contact pressure between the indenter and the layer is equal to the pressure acting at the interface between the layer and the elastic half-plane. This pressure is described by the periodic function $p(x)$. The tangential stresses are assumed to be zero or not affecting the normal pressure distribution. Then the displacement $u_2(x)$ of the boundary of the elastic half-plane is related to the pressure $p(x)$ acting on this boundary through the periodically located contact regions $[(2n+1)l/2-a, (2n+1)l/2+b], (n=0,\pm1,\pm2,\ldots)$ by the expression (Staierman, 1949):

$$u_2(x) = -\frac{2(1-\nu^2)}{\pi E_2}\int\limits_{l/2-a}^{l/2+b} p(x')\ln\left|2\sin\left[\frac{\pi(x'-x)}{l}\right]\right|dx', \tag{5.4}$$

where E_2 and ν are the elasticity modulus and Poisson's ratio of the half-plane, respectively.

Let the function $f(x)$ be smooth and have the form

$$f(x) = \frac{1}{2R}\left(x - \frac{l}{2}\right)^2, \quad x \in (0, l), \tag{5.5}$$

so the contact pressure $p(x)$ at the ends of the contact regions satisfies the condition

$$p\left(\frac{l}{2} - a\right) = p\left(\frac{l}{2} + b\right) = 0. \tag{5.6}$$

In virtue of the periodicity of the function $p(x)$, we have

$$p(x) = p(x + l). \tag{5.7}$$

In the contact regions, the condition of contact is satisfied, which for the rigid indenter has the form

$$\begin{aligned} &u_1(x) + u_2(x) = D - f(x), \\ &x \in [(2n+1)l/2 - a, \ (2n+1)l/2 + b], \quad (n = 0, \pm 1, \pm 2, \ldots), \end{aligned} \tag{5.8}$$

where D is the penetration of the punch into the two-layered foundation.

5.2.2 Method of solution

Relations (5.2, 5.4, 5.5, 5.8) with the periodicity conditions Eqs. (5.3) and (5.7) and boundary condition Eq. (5.6) form a complete system of equations for the unknown contact pressure. By eliminating the functions $u_i(x)$ $(i = 2, 3)$ from Eqs. (5.2), (5.4), and (5.8) and using Eq. (5.5), we obtain the equation that holds for all points of the contact region, i.e., for $x \in [l/2 - a, l/2 + b]$:

$$\begin{aligned} &-\frac{2(1-\nu^2)}{\pi E_2} \int_{l/2-a}^{l/2+b} p(x') \left[\ln\left| 2 \sin \frac{\pi(x'-x)}{l} \right| + \frac{\pi T_\varepsilon V}{l} \operatorname{ctg} \frac{\pi(x'-x)}{l} \right] dx' + \\ &+ \frac{h}{E_L} p(x) - \frac{h T_\sigma V}{E_L} p'(x) = D - \frac{1}{2R}\left(x - \frac{l}{2}\right)^2 + \frac{T_\varepsilon V}{R}\left(x - \frac{l}{2}\right). \end{aligned} \tag{5.9}$$

The solution of Eq. (5.2) in the regions where $p(x) = 0$, i.e., for $x \in [-l/2 + b, l/2 - a]$, is the function:

$$u_1 = u_0 \exp(x / T_\varepsilon V). \tag{5.10}$$

This function should satisfy the contact condition Eq. (5.8) for $x = l/2 - a$ and $x = l/2 + b$, as well as the periodicity condition Eq. (5.3), from which it follows that

$$u_0 \exp\left(\frac{b - l/2}{T_\varepsilon V}\right) = D - \frac{b^2}{2R} + \frac{2(1-\nu^2)}{\pi E_2} \int\limits_{l/2-a}^{l/2+b} p(x') \ln\left|2 \sin\frac{\pi(x' + l/2 - b)}{l}\right| dx'$$

$$u_0 \exp\left(\frac{l/2 - a}{T_\varepsilon V}\right) = D - \frac{a^2}{2R} + \frac{2(1-\nu^2)}{\pi E_2} \int\limits_{l/2-a}^{l/2+b} p(x') \ln\left|2 \sin\frac{\pi(x' - l/2 + a)}{l}\right| dx'. \tag{5.11}$$

By eliminating the constants D and u_0 from Eqs. (5.9) and (5.11), we obtain the following linear integral-differential equation for the dimensionless contact pressure $\widetilde{p}(\xi) = \frac{2(1-\nu^2)}{\pi E_2} p\left(\frac{b - a + l}{2} + \frac{a + b}{2}\xi\right)$ $(\xi \in [-1, 1])$:

$$\int\limits_{-1}^{1} \widetilde{p}(\xi') H(\xi, \xi') d\xi' - \frac{\beta_\varepsilon}{L} \widetilde{p}(\xi) + \frac{\beta_\varepsilon}{L \varsigma_0 \alpha_T} \widetilde{p}'(\xi) = G(\xi), \tag{5.12}$$

where the kernel $H(\xi,\xi')$ and the function $G(\xi)$ have the following form:

$$H(\xi,\xi') = \ln\left|2\sin\frac{\pi L}{2\tilde{l}}(\xi'-\xi)\right| + \frac{\pi L}{2\zeta_0\tilde{l}}\operatorname{ctg}\frac{\pi L}{2\tilde{l}}(\xi'-\xi)$$

$$+\frac{\ln\left|2\sin\frac{\pi L}{2\tilde{l}}(1-\xi')\right|\exp\left(2\zeta_0\left(\frac{\tilde{l}}{L}-1\right)\right)-\ln\left|2\sin\frac{\pi L}{2\tilde{l}}(1+\xi')\right|}{1-\exp\left(2\zeta_0\left(\frac{\tilde{l}}{L}-1\right)\right)}$$

$$G(\xi) = \frac{L}{2}\xi^2 + L\xi\left(\varepsilon-\frac{1}{\zeta_0}\right) - \frac{L\varepsilon}{\zeta_0} - \frac{L}{2}(1+2\varepsilon) + \frac{2\varepsilon L}{1-\exp\left(2\zeta_0\left(\frac{\tilde{l}}{L}-1\right)\right)} \tag{5.13}$$

In Eqs. (5.12) and (5.13), the dimensionless coordinate is introduced,

$$x = \frac{b-a+l}{2} + \frac{a+b}{2}\xi. \tag{5.14}$$

And the following dimensionless parameters are used:

$$L = \frac{a+b}{2R},\quad \varepsilon = \frac{b-a}{a+b},\quad \beta_\varepsilon = \frac{H\pi E_2}{2(1-\nu^2)RE_1},$$
$$\alpha_T = \frac{T_\varepsilon}{T_\sigma},\quad \zeta_0 = \frac{a+b}{2VT_\varepsilon},\quad \tilde{l} = \frac{l}{2R}. \tag{5.15}$$

These parameters depend on the mechanical characteristics of the viscoelastic layer (E_1, T_ε, T_σ), its thickness H, elastic characteristics of the half-plane (E_2 and ν), as well as on the curvature radius of an asperity R and distance between asperities l.

To determine the unknown size $(a+b)$ of the contact zone and the indentation D for the prescribed load P acting on each asperity, we use the equilibrium condition

$$\int_{-1}^{1}\tilde{p}(\xi)d\xi = \tilde{P},\quad \tilde{P} = \frac{2P(1-\nu^2)}{\pi E_2 R}, \tag{5.16}$$

and the condition of continuity for the pressure at the ends of the contact region in virtue of smoothness of the asperities (Eq. 5.6), which in the dimensionless form is written as

$$\widetilde{p}(-1) = \widetilde{p}(1) = 0. \tag{5.17}$$

The numerical solution of Eq. (5.12) taking into account Eqs. (5.16) and (5.17) was carried out by reducing them to the linear algebraic system of equations.

5.2.3 Analysis of the contact characteristic and friction force

The obtained solution of the contact problem is used to analyze the combined influence of the properties of the thin viscoelastic layer and the parameters of the indenter microrelief on the distribution of contact pressures and internal stresses in the elastic foundation, as well as on the size and shift of separate contact zones.

The properties of the surface layer modeled by the Kelvin solid are described by the dimensionless parameters α_T, ς_0, and β_ε (Eq. 5.15). The parameter α_T characterizes the viscous properties of the layer. The case $\alpha_T = 1$ corresponds to the elastic layer whose modulus of elasticity is equal to the long-term modulus E_1. The parameter ς_0 depends on the retardation time T_ε and sliding velocity V and is equal to the ratio of the time, during which an asperity passes the distance $(a + b)/2$, to the retardation time of the viscoelastic material. The parameter β_ε characterizes the relative thickness and elastic modulus of the surface layer. The case $\beta_\varepsilon \to +\infty$ corresponds to the model of a rigid indenter sliding over the viscoelastic layer bonded to the rigid base. The surface relief (waviness) is characterized by the dimensionless distance $\widetilde{l}$ between the asperities.

In Fig. 5.3, plots of the dimensionless contact pressures $\widetilde{p}$ are presented for various values of the parameters α_T and $\widetilde{l} = l/(2R)$. The results indicate that as α_T increases, the contact pressure distributions become more asymmetric. Decreasing the distance $l/(2R)$ between the asperities partly reduces this asymmetry.

Apart from asymmetric contact pressure, a shift of the contact region with respect to the axis of symmetry of each asperity takes place. The value of this shift depends on the distance between asperities (Fig. 5.4).

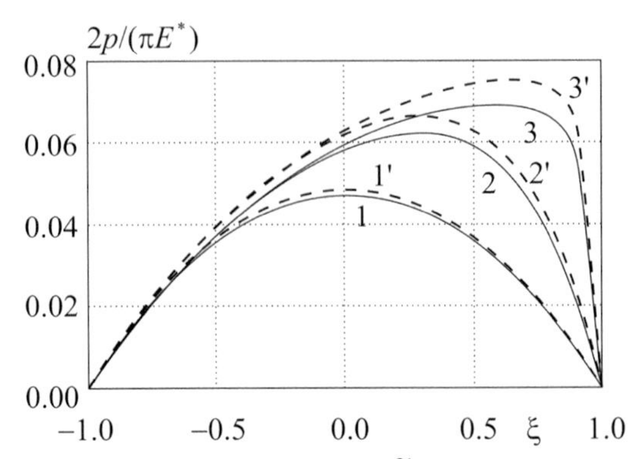

Figure 5.3 Contact pressure distribution for $\widetilde{P} = 1/15$, $\beta_\varepsilon = 10$, $T_\varepsilon V/R = 1$ and $\alpha_T = 1$ (1, 1′), $\alpha_T = 5$ (2, 2′), $\alpha_T = 50$ (3, 3′); $\widetilde{l} = 5$ (1, 2, 3), $\widetilde{l} = 1$ (1′, 2′, 3′).

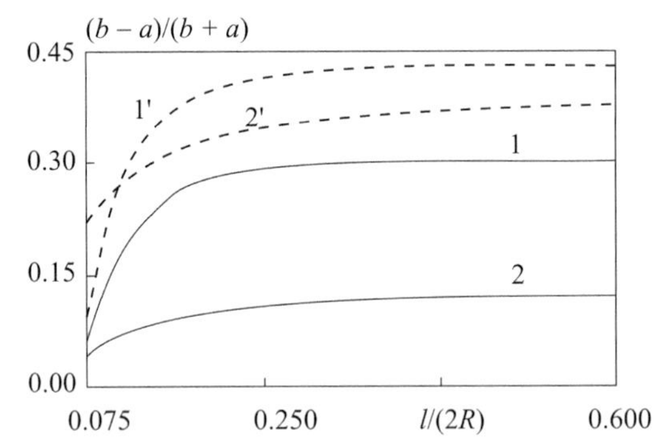

Figure 5.4 Shift of the contact region as a function of distance between asperities for $\alpha_T = 50$, $\widetilde{P} = 1/15$ and $T_\varepsilon V/R = 1$ (1, 1′), $T_\varepsilon V/R = 10$ (2, 2′); $\beta_\varepsilon = 10$ (solid lines) and $\beta_\varepsilon \to +\infty$ (dashed lines).

The results of calculations show that the effect of the distance between neighboring asperities on the contact characteristics becomes significant only for sufficiently small values of the parameter $l/(2R)$ $(l/(2R) < 0,2)$. The comparison of curves 1, 2 and 1′, 2′ in Fig. 5.4 (the latter pair of curves is constructed for the rigid substrate) shows that taking into account elasticity of the substrate decreases the shift of the contact region, which is more noticeable for large values of the parameter $T_\varepsilon V/R$, i.e., for high sliding velocities.

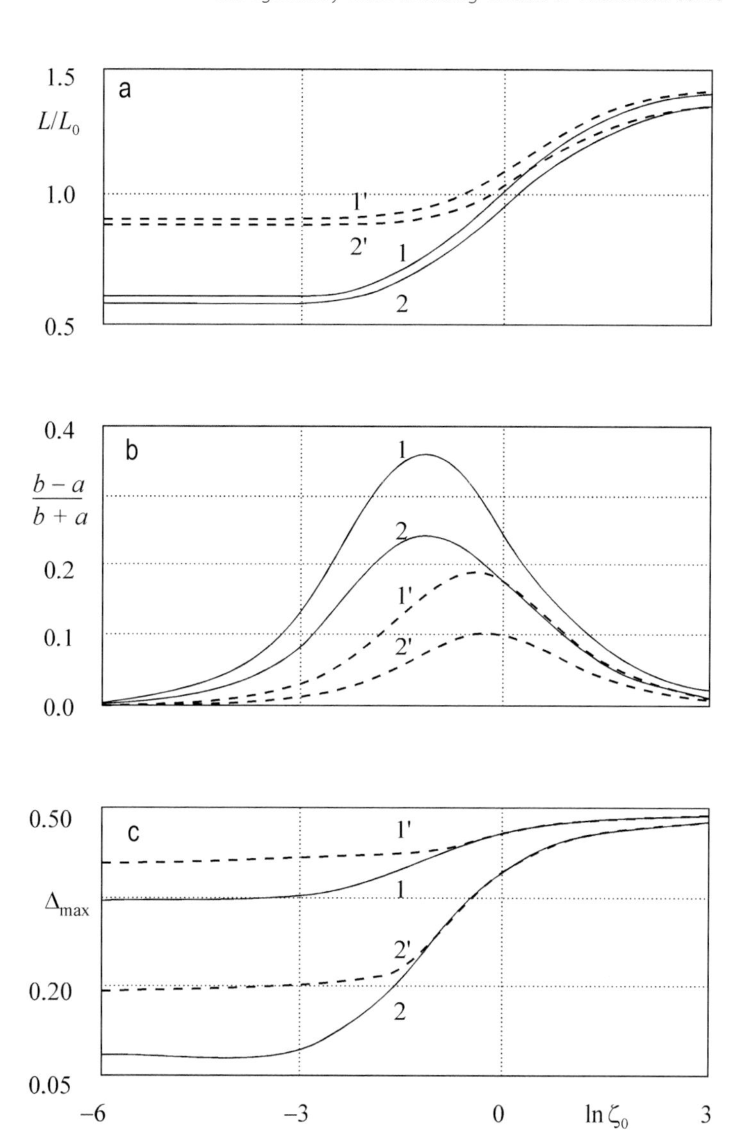

Figure 5.5 Width (a) and shift (b) of the contact region and its maximum indentation (c) versus the parameter ς_0 for $\widetilde{P} = 1/15$, $\beta_\varepsilon = 10$ and $\widetilde{l} = 5$ (1, 1′), $\widetilde{l} = 1$ (2, 2′); $\alpha_T = 50$ (1, 2), $\alpha_T = 5$ (1′, 2′).

Fig. 5.5 represents plots of the dimensionless width L/L_0 of the contact region for an asperity, its shift ε, and maximum indentation $\Delta_{\max}$ of the asperity into the viscoelastic layer as functions of the parameter ς_0 for various values of α_T and $\tilde{l}$. For the calculation of the value $\Delta_{\max}$, we use the following relation obtained from Eqs. (5.4), (5.8), and (5.11):

$$\Delta_{\max} = \max_{x \in (-a,b)} \frac{u_1}{R}$$
$$= L \int_{-1}^{1} \tilde{p}(\xi) \left[H(0,\xi) - \frac{\pi L}{2\tilde{l}\,\zeta_0} \operatorname{ctg} \frac{\pi L}{2\tilde{l}} \xi \right] d\xi + \frac{L^2(\varepsilon+1)^2}{2}$$
$$+ \frac{2\varepsilon L^2}{\exp\left(2\zeta_0\left(\frac{\tilde{l}}{L} - 1\right)\right) - 1}. \quad (5.18)$$

The results of calculations allow us to conclude that the width of the contact region and maximum indentation into the viscoelastic layer are restricted by their limit values for $\varsigma_0 \to 0$ and $\varsigma_0 \to \infty$. The case $\varsigma_0 \to \infty$, i.e., $V \to 0$, corresponds to the elastic layer with the modulus E_L bonded to the elastic half-plane. In the case $\varsigma_0 \to 0$, i.e., $V \to \infty$, deformation of the surface layer is defined by its instantaneous modulus of elasticity $\alpha_T E_1$. Note that as the sliding velocity V increases (the parameter ς_0 decreases), the contact width and indentation of the indenter decrease (floating effect). The shift of the contact region attains its maximum for a certain value of ς_0, and it tends to zero as $\varsigma_0 \to 0$ and $\varsigma_0 \to \infty$.

Shift of the contact region with respect to the axis of symmetry of an asperity, as well as the asymmetry of the contact pressure distribution, cause a tangential force T acting on each asperity in the direction opposite to sliding. This force is called the deformation (hysteretic) friction force. The dimensionless value of this force can be calculated as

$$\tilde{T} = \frac{2(1-\nu^2)}{\pi R^2 E_2} \int_{-a}^{b} x p(x) dx. \quad (5.19)$$

The deformation component μ of the coefficient of friction is calculated taking into account Eqs. (5.14–5.16) and (5.19) as

$$\mu = \frac{\tilde{T}}{\tilde{P}} = \frac{L^2}{\tilde{P}} \int_{-1}^{1} (\varepsilon + \xi)\tilde{p}(\xi) d\xi. \quad (5.20)$$

Formula (5.20) allows one to analyze the dependence of the hysteretic component of the coefficient of friction on the parameters of the indenter relief, rheological characteristics of the surface layer, and sliding velocity. In particular, it is shown (Fig. 5.6) that the hysteretic component μ of the coefficient of friction is a nonmonotonic function of the parameter $\varsigma_0 = (a+b)/(2VT_\varepsilon)$, and it tends to zero as $\varsigma_0 \to 0$ and $\varsigma_0 \to \infty$. Those limit cases correspond to purely elastic formulations of the problem.

The results presented in Fig. 5.6 are calculated for the following values of the dimensionless parameters: $\beta_\varepsilon = 10$, $\widetilde{P} = 1/15$ and $T_\varepsilon/T_\sigma = 50$ (solid lines), $T_\varepsilon/T_\sigma = 5$ (dashed lines). Curves 1 and 1′ are constructed for $l/(2R) = 5$, whiles curves 2 and 2′ for $l/(2R) = 1$. The comparison of the curves shows that an increase in the parameter T_ε/T_σ, which characterizes the viscosity of the surface layer, as well as a decrease in the distance between asperities lead to a growth of the coefficient of friction. But if the length $(l-a-b)$ of the interval between the contact regions is sufficiently small and the inequality $(l-a-b)/V << T_\varepsilon$ is satisfied, the layer does not have enough time to restore its shape, which leads to a decrease in the contact area and its shift and, consequently, to a decrease in the coefficient of friction. This case was investigated in detail previously (Goryacheva, 1998). It was also shown there that taking into account elasticity of the indenter and substrate leads to a decrease in the shift of the contact area, and, accordingly, in the coefficient of friction, which is particularly so for large values of the parameter $T_\varepsilon V/R$, i.e., at high sliding velocities.

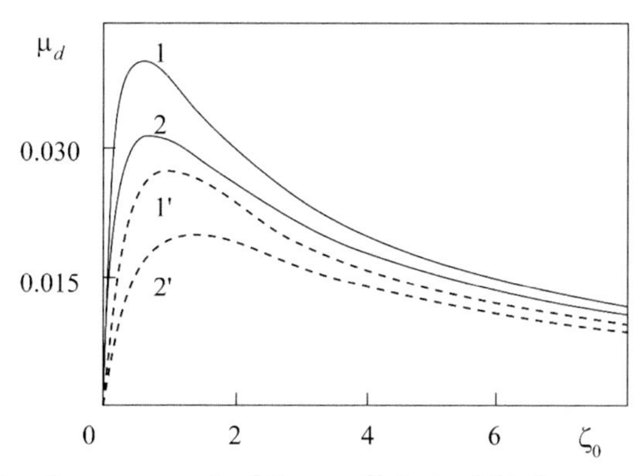

Figure 5.6 Hysteretic component of the coefficient of friction as a function of the parameter ζ_0.

5.3 Sliding contact of a punch with periodic microgeometry over the viscoelastic half-plane (2-D analysis)

In this section, an analytical model that takes into account both periodic shape of a punch and imperfect elasticity of a counterbody modeled by an isotropic viscoelastic solid is developed to study the combined effect of the sliding velocity and the microgeometry parameters on the contact pressure distribution and the friction force in sliding contact.

5.3.1 Constitutive relations for the viscoelastic half-plane

Relations between the components of strains $\varepsilon_{x^0}, \varepsilon_{y^0}, \gamma_{x^0y^0}$ and stresses $\sigma_{x^0}, \sigma_{y^0}, \tau_{x^0y^0}$ in an isotropic viscoelastic solid in the case of plane strain are taken in the form (Goryacheva, 1998):

$$
\begin{aligned}
\varepsilon_{x^0} + T_\varepsilon \frac{\partial \varepsilon_{x^0}}{\partial t} &= \frac{1-\nu^2}{E}\left(\sigma_{x^0} + T_\sigma \frac{\partial \sigma_{x^0}}{\partial t}\right) - \frac{\nu(1+\nu)}{E}\left(\sigma_{y^0} + T_\sigma \frac{\partial \sigma_{y^0}}{\partial t}\right) \\
\varepsilon_{y^0} + T_\varepsilon \frac{\partial \varepsilon_{y^0}}{\partial t} &= \frac{1-\nu^2}{E}\left(\sigma_{y^0} + T_\sigma \frac{\partial \sigma_{y^0}}{\partial t}\right) - \frac{\nu(1+\nu)}{E}\left(\sigma_{x^0} + T_\sigma \frac{\partial \sigma_{x^0}}{\partial t}\right) \\
\gamma_{x^0y^0} + T_\varepsilon \frac{\partial \gamma_{x^0y^0}}{\partial t} &= \frac{1+\nu}{E}\left(\tau_{x^0y^0} + T_\sigma \frac{\partial \tau_{x^0y^0}}{\partial t}\right)
\end{aligned}
\tag{5.21}
$$

Here, T_ε and T_σ characterize viscous properties of the solid ($T_\varepsilon > T_\sigma$), E and ν are the long-term Young modulus and Poisson's ratio, and $E_I = T_\varepsilon E / T_\sigma$ is the instantaneous modulus. Relations Eq. (5.21) are a 2-D version of the 1-D Kelvin model given by Eq. (5.1).

When the punch moves over the boundary of the viscoelastic half-plane with a constant velocity V, the stress and strain distributions in the half-plane can be considered steady with respect to the system of coordinates, $x = x^0 - Vt$, $y = y^0$, that is moving with the punch. In this system of coordinates, the stresses and displacements do not depend explicitly on time; they are functions of only the coordinates (x, y). We introduce the fictitious stresses according to the following formulas:

$$
\begin{aligned}
\varepsilon^0_{ij} + T_\varepsilon \frac{\partial \varepsilon^0_{ij}}{\partial t} &= \varepsilon_{ij} - T_\varepsilon V \frac{\partial \varepsilon_{ij}}{\partial x} = \varepsilon^*_{ij} \\
\sigma^0_{ij} + T_\sigma \frac{\partial \sigma^0_{ij}}{\partial t} &= \sigma_{ij} - T_\sigma V \frac{\partial \sigma_{ij}}{\partial x} = \sigma^*_{ij}
\end{aligned}
\tag{5.22}
$$

The fictitious stresses satisfy the equations that are equivalent to the equations of equilibrium and compatibility and the Hooke's law for an isotropic elastic solid.

From Eq. (5.22) it follows that the fictitious displacement $u^*(x)$ and fictitious pressure $p^*(x)$ are related to the real displacement $u(x)$ of the deformed boundary of the viscoelastic half-plane in the y-direction, which is perpendicular to the boundary of the half-plane, and to the normal pressure $p(x)$ acting on this boundary, respectively, by the expressions

$$\begin{aligned} u^*(x) &= u(x) - T_\varepsilon V u'(x), \\ p^*(x) &= p(x) - T_\sigma V p'(x) \end{aligned} \tag{5.23}$$

(prime denotes derivative with respect to x).

5.3.2 Saturated periodic contact of the elastic half-plane

Consider a 2-D quasistatic problem for a punch with a periodic relief $f(x) = f(x + l)$ sliding with a constant velocity V over the boundary of the viscoelastic half-plane in the x-direction. At the boundary of the half-plane, the condition $u'(x) = f'(x)$ is satisfied in the contact regions, and $p(x) = 0$ is satisfied outside these regions. Tangential stresses at the boundary are assumed to be zero.

As the punch slides over the viscoelastic half-plane, asymmetry of the contact stress distribution causes the friction force. The linear friction force T (applied to a unit length of the indenter) is directed along the x-axis and can be calculated as

$$T = \int_0^l p(x) f'(x) dx = \int_0^l p(x) u'(x) dx. \tag{5.24}$$

Relation (5.24) is obtained under the assumption that the amplitude h of asperities of the relief is much smaller than its period l ($h << l$).

For the punch to move steadily, it is necessary that it is acted upon by a tangential force balancing the friction force. Besides, the punch is pressed to the viscoelastic base by the normal linear force P that acts on a period and balances the contact pressure $p(x)$:

$$P = \int_0^l p(x) dx. \tag{5.25}$$

To establish the relation between the displacement $u(x)$ of the boundary of the viscoelastic half-plane and the pressure $p(x)$ applied to this boundary, and also to determine the friction force T, we first introduce the following dimensionless values and functions:

$$\tilde{x} = \frac{x}{l}, \quad \tilde{t} = \frac{t}{l}, \quad \tilde{p}\left(\tilde{x}\right) = \frac{2(1-\nu^2)}{\pi E} p\left(\tilde{x}\, l\right), \quad \tilde{p}*\left(\tilde{x}\right) = \frac{2(1-\nu^2)}{\pi E} p^*\left(\tilde{x}\, l\right)$$

$$\tilde{u}\left(\tilde{x}\right) = \frac{u\left(\tilde{x}\, l\right)}{l}, \quad \tilde{u}*\left(\tilde{x}\right) = \frac{u^*\left(\tilde{x}\, l\right)}{l}, \quad \tilde{P} = \frac{2(1-\nu^2)}{\pi El} P = \int_0^1 \tilde{p}\left(\tilde{x}\right) d\tilde{x}$$

$$\alpha = \frac{T_\varepsilon}{T_\sigma}, \quad \beta = \frac{h}{l}, \quad \zeta = \frac{l}{2VT_\varepsilon}, \quad \tilde{T} = \frac{2(1-\nu^2)}{\pi El} T = \int_0^1 \tilde{p}\left(\tilde{x}\right) \tilde{u}'_{\tilde{x}}\left(\tilde{x}\right) d\tilde{x} \tag{5.26}$$

In this notation, Eq. (5.23) has the form (here and in what follows, the tilde sign is omitted):

$$u^*(x) = u(x) - \frac{u'(x)}{2\zeta}, \quad p^*(x) = p(x) - \frac{p'(x)}{2\alpha\zeta} \tag{5.27}$$

From the solution of the problem of elasticity, it follows that if the function $p^*(x)$ at the boundary of the elastic half-plane is prescribed, the fictitious displacement $u^*(x)$ of the boundary is related to the fictitious pressure $p^*(x)$ by the following integral operator (Staierman, 1949):

$$\int_0^1 p^*(t) \ln|2 \sin \pi(t-x)| dt = u^*(x). \tag{5.28}$$

Hence, the derivative of the fictitious displacement $u^{*\prime}(x)$ is the Hilbert transform of the fictitious pressure $p^*(x)$,

$$-\pi \int_0^1 p^*(t) \mathrm{ctg} \pi(t-x) dt = u^{*\prime}(x), \tag{5.29}$$

so the function $p^*(x)$ can be expressed through the function $u^{*\prime}(x)$ by using the inverse Hilbert transformation:

$$p^*(x) = \int_0^1 p^*(t)dt + \frac{1}{\pi}\int_0^1 \text{ctg}\pi(t-x)u^{*\prime}(t)dt. \tag{5.30}$$

If the $u^*(x)$ is expanded into the Fourier series as

$$u^*(x) = \frac{\alpha_0^+}{2} + \sum_{k=1}^{\infty}\left(\alpha_k^+ \cos 2\,\pi kx + \alpha_k^- \sin 2\,\pi kx\right), \tag{5.31}$$

then by substituting this series into Eq. (5.30) and calculating the integrals, we obtain

$$p^*(x) = \int_0^1 p^*(t)dt - 2\sum_{k=1}^{\infty} k\left(\alpha_k^+ \cos 2\,\pi kx + \alpha_k^- \sin 2\,\pi kx\right). \tag{5.32}$$

Based on Eqs. (5.27), (5.31), and (5.32), we find relation between the coefficients of the Fourier expansion of the displacement function $u(x)$ and those of the pressure function $p(x)$. Let us assume the following (note that the function $u(x)$ is defined up to a constant, so the initial term is insignificant):

$$u(x) = \frac{a_0^+}{2} + \sum_{k=1}^{\infty}\left(a_k^+ \cos 2\,\pi kx + a_k^- \sin 2\,\pi kx\right) \tag{5.33}$$

and

$$p(x) = \int_0^1 p(t)dt + \sum_{k=1}^{\infty}\left(A_k^+ \cos 2\,\pi kx + A_k^- \sin 2\,\pi kx\right)$$
$$\left[\int_0^1 p(t)dt = \int_0^1 p^*(t)dt = P\right]. \tag{5.34}$$

Here, equations in the square brackets follow from relations (5.26) and (5.27) and from the conditions of periodicity of the functions. Based on expansions (5.33) and (5.34), from relations (5.27), (5.31), and (5.32), we get the following expression relating the expansion coefficients:

$$A_k^{\pm} \mp \frac{\pi k}{\alpha\zeta}A_k^{\mp} = -2k\left(a_k^{\pm} \mp \frac{\pi k}{\zeta}a_k^{\mp}\right). \tag{5.35}$$

Here, one should take only upper or lower "plus" and "minus" signs. From Eq. (5.35) we obtain the coefficients $A_k^{\pm}$ expressed via the coefficients $a_k^{\pm}$ (Goryacheva and Goryachev, 2016):

$$A_k^{\pm} = -2k \frac{\left(1 + \frac{1}{\alpha}\left(\frac{\pi k}{\zeta}\right)^2\right) a_k^{\pm} \mp \frac{\pi k}{\zeta}\left(1 - \frac{1}{\alpha}\right) a_k^{\mp}}{1 + \left(\frac{\pi k}{\alpha \zeta}\right)^2}. \quad (5.36)$$

From Eq. (5.36) one can also obtain the expressions for the coefficients $a_k^{\pm}$ through the coefficients $A_k^{\pm}$ (Goryacheva and Goryachev, 2016):

$$a_k^{\pm} = -\frac{1}{2k} \frac{\left(1 + \frac{1}{\alpha}\left(\frac{\pi k}{\zeta}\right)^2\right) A_k^{\pm} \pm \frac{\pi k}{\zeta}\left(1 - \frac{1}{\alpha}\right) A_k^{\pm}}{1 + \left(\frac{\pi k}{\zeta}\right)^2}. \quad (5.37)$$

Note that in Eqs. (5.35)–(5.37), $k = 1, 2, 3, \ldots$

If the contact pressure distribution over a period or the punch shape are prescribed, for the case of saturated contact, from relations (5.34) and (5.35)–(5.37), we obtain the expression for the dimensionless tangential friction force T through the expansion coefficients $a_k^{\pm}$ of the function $u(x)$ or/and through the expansion coefficients $A_k^{\pm}$ of the function $p(x)$ (Goryacheva and Goryachev, 2016):

$$T = \int_0^1 p(x) u'(x) dx = \pi \sum_{k=1}^{\infty} k\left(A_k^{+} a_k^{-} - a_k^{+} A_k^{-}\right)$$
$$= \frac{2\pi^2}{\zeta}\left(1 - \frac{1}{\alpha}\right) \sum_{k=1}^{\infty} \frac{k^3\left(a_k^{+2} + a_k^{-2}\right)}{1 + \left(\frac{\pi k}{\alpha \zeta}\right)^2} = \frac{\pi^2}{2\zeta}\left(1 - \frac{1}{\alpha}\right) \sum_{k=1}^{\infty} \frac{k\left(A_k^{+2} + A_k^{-2}\right)}{1 + \left(\frac{\pi k}{\zeta}\right)^2}. \quad (5.38)$$

5.3.3 Saturated contact characteristics for the viscoelastic half-plane and punches with various relief shape

Wavy profile is described by the function

$$f(x) = h \cos^2 \frac{\pi x}{l}. \quad (5.39)$$

Let the periodic profile (Fig. 5.7a) be specified by the function Eq. (5.39). The saturated contact is assumed between the viscoelastic half-plane and the punch sliding with a prescribed velocity V; i.e., the contact condition $u(x) = f(x) + D$ is satisfied for $-\infty < x < +\infty$.

The constant D characterizes the indentation of the punch under a prescribed load P. This constant does not influence the solution of the problem. Thus, for the elastic displacement of the boundary, from the contact condition, we obtain the following expression in the dimensionless form:

$$u(x) = \frac{\beta}{2}(1 + \cos 2\pi x). \tag{5.40}$$

Then, in accordance with Eqs. (5.34) and (5.36), the dimensionless contact pressure is determined as

$$p(x) = \int_0^1 p(t)dt - \beta \frac{\left(1 + \frac{1}{\alpha}\left(\frac{\pi}{\zeta}\right)^2\right)\cos 2\pi x + \frac{\pi}{\zeta}\left(1 - \frac{1}{\alpha}\right)\sin 2\pi x}{1 + \left(\frac{\pi}{\alpha\zeta}\right)^2}, \tag{5.41}$$

which taking into account the notations Eq. (5.26) can be represented by the expressions

$$p(x) = P - P_0 \cos(2\pi x - \phi);\ P_0 = \beta\sqrt{\frac{1 + \left(\frac{\pi}{\zeta}\right)^2}{1 + \left(\frac{\pi}{\alpha\zeta}\right)^2}},$$
$$\phi = \operatorname{arctg}\frac{\frac{\pi}{\zeta}\left(1 - \frac{1}{\alpha}\right)}{1 + \frac{1}{\alpha}\left(\frac{\pi}{\zeta}\right)^2}. \tag{5.42}$$

On the basis of Eq. (5.38), the dimensionless tangential force is calculated as

$$T = \frac{\pi^2\beta^2}{2\zeta} \cdot \frac{1 - \frac{1}{\alpha}}{1 + \frac{1}{\alpha^2}\left(\frac{\pi}{\zeta}\right)^2}. \tag{5.43}$$

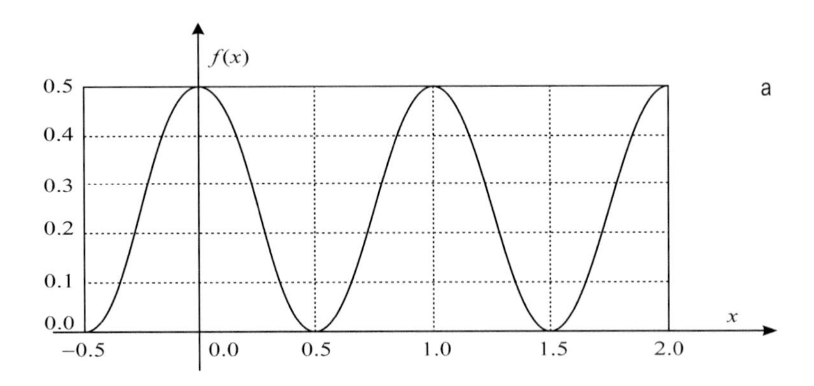

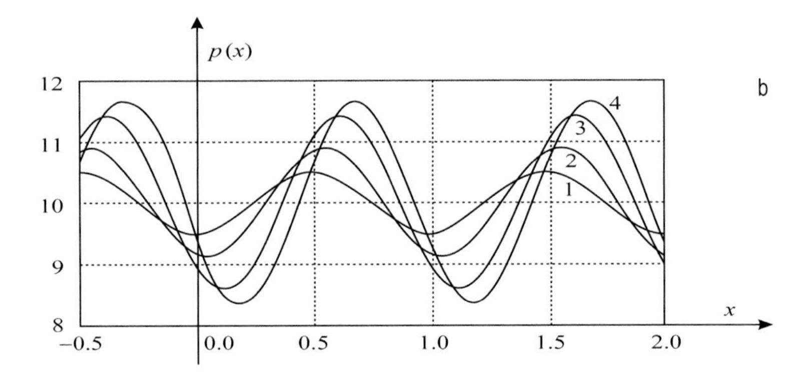

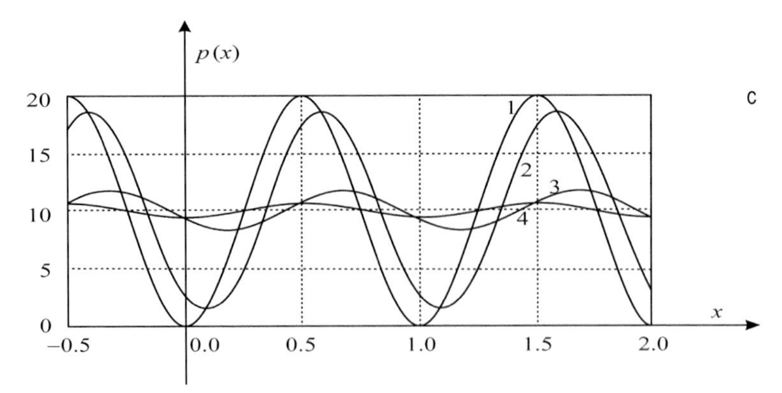

Figure 5.7 The punch shape (a) and the pressure distribution under this punch for $\beta = 1/2$, $P = 10$, and various values of the parameter α ($\varsigma = 1$): $\alpha = 1$ (1), $\alpha = 2$ (2), $\alpha = 5$ (3), $\alpha = 20$ (4) (b) and the parameter ς ($\alpha = 20$): $\varsigma = 10^{-3}$ (1), $\varsigma = 10^{-1}$ (2), $\varsigma = 1$ (3), $\varsigma = 10^{3}$ (4) (c).

From Eq. (5.43), it follows that the saturated contact occurs only when $P > P_0$.

In Fig. 5.7b and c, the pressure distributions are presented for $\beta = 1/2$, $P = 10$, and various values of the parameters ζ and α. In Fig. 5.7b, the pressure distributions are calculated for $\varsigma = 1$ and the values of the parameter α are $\alpha = 1$ (curve 1), $\alpha = 2$ (curve 2), $\alpha = 5$ (curve 3), $\alpha = 20$ (curve 4). From the analysis of Eq. (5.42) and the results of calculations, it follows that for $\alpha = 1$ (the case of elastic half-plane, $E_I = E$), the pressures are distributed symmetrically with respect to the axis of symmetry of an asperity, with the amplitude of pressures being the smallest and equal to 2β. As the viscosity of the half-plane increases (the parameter α increases), the maximum contact pressure shifts from the axis of symmetry of an asperity in the direction of its motion. An increase in the parameter α leads to an increase in this shift and in the amplitude of pressure.

Fig. 5.7c depicts the pressure distributions calculated by Eq. (5.42) for $\alpha = 20$ and the following values of the parameter ς, which is inversely proportional to the sliding velocity: $\varsigma = 10^{-3}$ (curve 1), $\varsigma = 10^{-1}$ (curve 2), $\varsigma = 1$ (curve 3), $\varsigma = 10^3$ (curve 4). The function of pressure is periodic; its amplitude values increase with increasing velocity (decreasing the parameter ς). At high and low sliding velocities, the pressure distribution becomes symmetric with respect to the axis of symmetry of an asperity. From Eq. (5.42) it follows that the maximum shift of the pressure distribution from the axis takes place at $\varsigma = \pi/\sqrt{\alpha}$ ($\alpha > 1$).

In Fig. 5.8, graphs of the deformation friction force (5.43) as a function of the parameter ς are presented for $\beta = 1/2$ and various values of α.

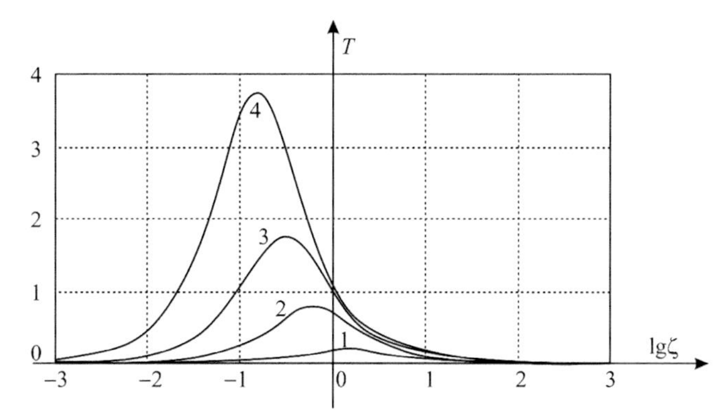

Figure 5.8 Hysteretic component of the friction force versus the parameter ς for $\beta = 1/2$ and $\alpha = 2$ (1), $\alpha = 5$ (2), $\alpha = 10$ (3), and $\alpha = 20$ (4).

The deformation (hysteretic) component of the friction force is a non-monotone function of time: it tends to zero as sliding velocity tends to zero or infinity ($\zeta \to \infty$ or $\zeta \to 0$). As was mentioned in Section 5.2, these limit cases correspond to elastic solutions with the long-term and instantaneous moduli of elasticity, respectively, for which the pressure is distributed symmetrically with respect to the axis of symmetry of an asperity. It follows from Eq. (5.57) that the maximum value of the friction force $T_{\max} = \frac{\pi\beta^2(\alpha-1)}{4}$ is attained for $\zeta = \frac{\pi}{\alpha}$, i.e., $\lg\zeta = \lg\pi - \lg\alpha$. The friction force increases as the parameter α grows, i.e., as the viscosity of the half-plane increases. From Eq. (5.43) it also follows that the friction force increases with an increase in the asperity height ($T \sim \beta^2$).

Wavy profile is described by the function

$$f(x) = h\cos^4\frac{\pi x}{l}. \tag{5.44}$$

Now let the periodic profile (Fig. 5.9a) be defined by the function Eq. (5.44), and let the saturated contact take place between the sliding profile and the viscoelastic half-plane. Then, in Eq. (5.33) we have $\frac{a_0^+}{2} = \frac{5\beta}{8}$, $a_1^+ = \frac{\beta}{2}$, $a_2^+ = \frac{\beta}{8}$, and the function $u(x)$ in the dimensionless coordinates has the form:

$$u(x) = \beta\left(\frac{3}{8} + \frac{1}{2}\cos 2\pi x + \frac{1}{8}\cos 4\pi x\right). \tag{5.45}$$

In this case, from Eq. (5.34) we obtain the dimensionless contact pressure:

$$p(x) = P + A_1^+\cos 2\pi x + A_1^-\sin 2\pi x + A_2^+\cos 4\pi x + A_2^-\sin 4\pi x. \tag{5.46}$$

Here, the dimensionless normal force P is determined from Eq. (5.34), whereas the expansion coefficients are given by Eq. (5.36):

$$\begin{array}{ll} A_1^+ = -\beta\dfrac{1+\dfrac{1}{\alpha}\left(\dfrac{\pi}{\zeta}\right)^2}{1+\left(\dfrac{\pi}{\alpha\zeta}\right)^2} & A_1^- = -\beta\dfrac{\dfrac{\pi}{\zeta}\left(1-\dfrac{1}{\alpha}\right)}{1+\left(\dfrac{\pi}{\alpha\zeta}\right)^2} \\ A_2^+ = -\dfrac{\beta}{2}\cdot\dfrac{1+\dfrac{4}{\alpha}\left(\dfrac{\pi}{\zeta}\right)^2}{1+4\left(\dfrac{\pi}{\alpha\zeta}\right)^2} & A_2^- = -\beta\dfrac{\dfrac{\pi}{\zeta}\left(1-\dfrac{1}{\alpha}\right)}{1+4\left(\dfrac{\pi}{\alpha\zeta}\right)^2} \end{array}. \tag{5.47}$$

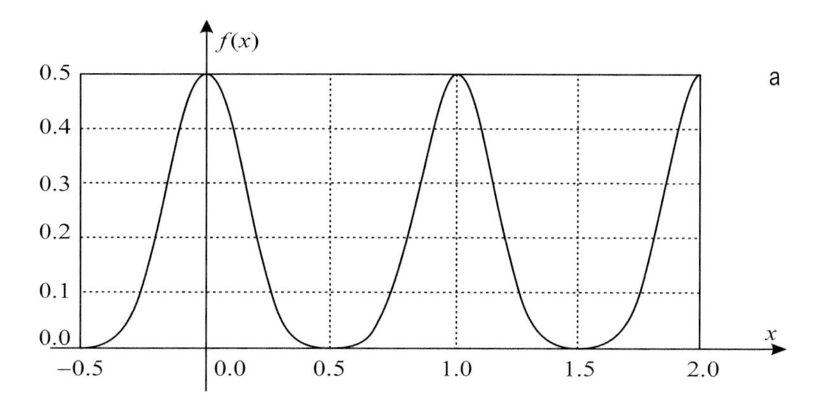

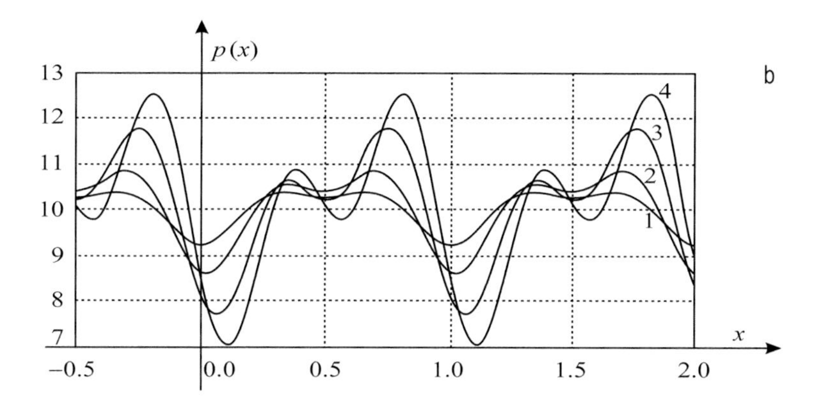

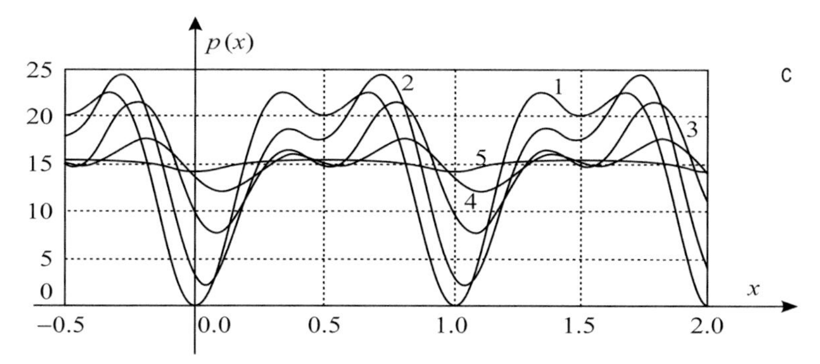

Figure 5.9 Punch shape (a) and pressure distribution under this punch for $\beta = 1/2$ and various values of the parameter α ($\varsigma = 1$, $P = 10$): $\alpha = 1$ (1), $\alpha = 2$ (2), $\alpha = 5$ (3), $\alpha = 20$ (4) (b) and parameter ς ($\alpha = 20$ $P = 15$): $\varsigma = 10^{-3}$ (1), $\varsigma = 10^{-1}$ (2), $\varsigma = 10^{-1/2}$ (3), $\varsigma = 1$ (4) $\varsigma = 10^{3}$ (5) (c).

From Eq. (5.46) it follows that for the saturated contact to exist, it suffices that the following condition is satisfied: $P \geq \sqrt{(A_1^+)^2 + (A_1^-)^2} + \sqrt{(A_2^+)^2 + (A_2^-)^2}$.

The dimensionless tangential force T is determined by the following expression obtained from Eqs. (5.38) and (5.35):

$$T = \frac{3\pi^2\beta^2}{4\zeta}\left(1 - \frac{1}{\alpha}\right)\frac{1 + 3\left(\frac{\pi}{\alpha\zeta}\right)^2}{\left(1 + \left(\frac{\pi}{\alpha\zeta}\right)^2\right)\left(1 + 4\left(\frac{\pi}{\alpha\zeta}\right)^2\right)}. \tag{5.48}$$

In Fig. 5.9b, the pressure distributions are presented for $\beta = 1/2$, $P = 10$, $\varsigma = 1$, and various values of the parameter α. From the results of calculations, it follows that for $\alpha = 1$ (the case of elastic half-plane, $E_I = E$), the pressures are distributed symmetrically with respect to the axis of symmetry of an asperity. In a period, there are two equal maximums of the function $p(x)$, which are located on an asperity. As the viscosity of the half-plane increases (the parameter α increases), the pressures increase on an asperity and decrease in a valley between asperities. Two maximums of the pressure function $p(x)$ given by Eq. (5.46) still take place, the first maximum (on the oncoming part of an asperity) being higher than the second one. Also, the points of local maximums shift in the direction of sliding of the punch. The values of the shift and the amplitude values of the pressure increase as the parameter α increases.

Fig. 5.9c illustrates the pressure distributions under the punch that are calculated by formula (5.46) for $\alpha = 20$, $\beta = 1/2$, $P = 15$, and various values of the parameter ς, which is inversely proportional to the sliding velocity of the punch (see relations Eq. 5.26). At high and low velocities, the pressure distributions are symmetrical with respect to the axis of symmetry of an asperity, the pressure amplitude increasing as the velocity increases. For intermediate values of the parameter ς, there are two local maximums of the function $p(x)$ on an asperity, with the maximum on the ongoing part of an asperity being higher than the second maximum. The difference between the values of these two maximums depends on the sliding velocity of the punch.

The hysteretic component of the friction force is presented in Fig. 5.10 as a function of the parameter ς for various values of the parameter α.

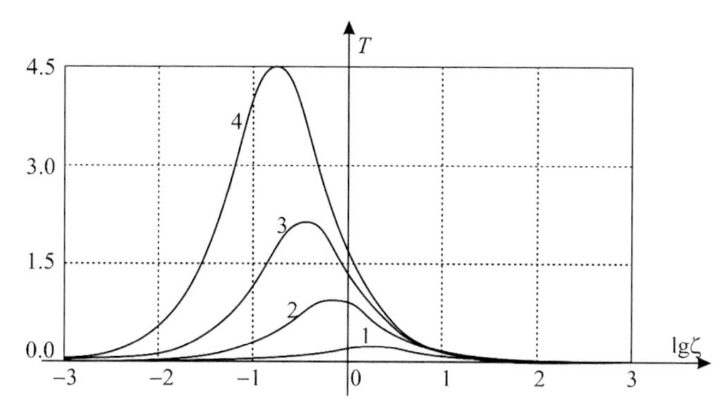

Figure 5.10 Hysteretic friction force as a function of the parameter ς for $\beta = 1/2$ and $\alpha = 2$ (1), $\alpha = 5$ (2), $\alpha = 10$ (3), and $\alpha = 20$ (4).

The dependencies of the hysteretic friction force on the parameters ς, α, $\varkappa$, and β, calculated by formula (5.48), are similar to those presented in Fig. 5.8 for the different punch shape. However, the flatter shape of an asperity top (Fig. 5.9a) leads to higher values of the friction force for the same values of the velocity, asperity height, and viscoelastic properties of the half-plane.

It is proved (Goryacheva ang Goryachev, 2016) that saturated contact does not exist for the periodic profile of sawtooth shape, which is specified in the dimensionless form as $u(x) = \beta|2x-1|, \quad x \in [0, 1]$.

Note that relations Eq. (5.37) also allow us, based on a given pressure distribution $p(x)$, to determine the shape of the deformed surface $u(x)$ (up to a constant) and, hence, to determine the shape of a punch that causes this pressure distribution (Goryacheva and Goryachev, 2016).

5.4 Sliding contact of a rigid cylinder over the viscoelastic layer with periodic surface microgeometry

In this section, an analytical solution of the contact problem for a cylinder sliding over the wavy viscoelastic base is constructed under the assumption that the plane of wave section is perpendicular to the cylinder generatrix.

5.4.1 Problem formulation

Consider sliding of a cylindrical punch of a radius R over a viscoelastic layer lying on a rigid base. The punch slides with a constant velocity V in the x-direction (Fig. 5.11). In the moving coordinate system attached to the

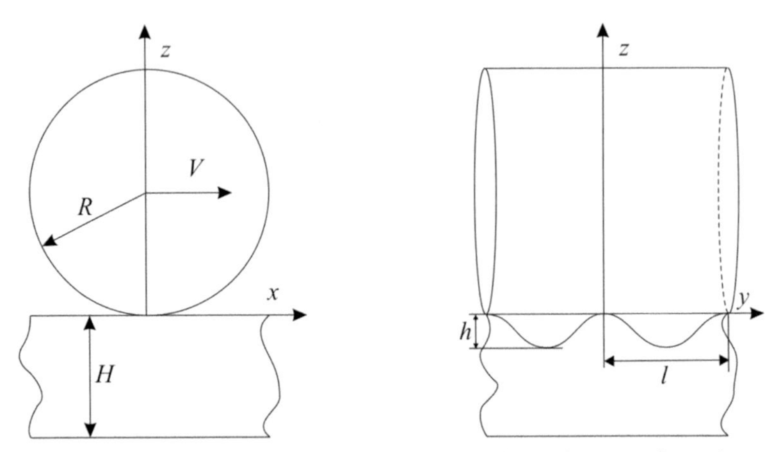

Figure 5.11 Scheme of contact between a cylinder and a viscoelastic layer.

cylinder, the contacting part of its surface is described by the function $z = f(x)$. The problem is considered in quasi-static formulation. The cylinder is acted upon by the constant linear force P/R directed opposite to the z-axis.

The viscoelastic layer has a thickness H, and its surface has a wavy relief described by the periodic function $\phi(y)$ with the period l and amplitude h. Note that $h \ll H$. Mechanical properties of the viscoelastic layer are described by the Kelvin model Eq. (5.1). In the moving system of coordinates (x, y, z), the relation between the layer displacement $u(x, y)$ in the direction opposite to the z-axis and the contact pressure $p(x, y)$ is

$$u - T_\varepsilon V \frac{\partial u}{\partial x} = \frac{H}{E}\left(p - T_\sigma V \frac{\partial p}{\partial x}\right). \tag{5.49}$$

Here, T_ε and T_σ are the retardation and relaxation times, respectively, and E is the long-term modulus of the viscoelastic layer.

The condition of contact between the punch and the viscoelastic layer over the region ω has the form:

$$u(x, y) = \Delta(y) - f(x); \quad (x, y) \in \omega, \tag{5.50}$$

where

$$\Delta(y) = D - \phi(y). \tag{5.51}$$

Here, D is the indentation of the punch into the viscoelastic layer, and $\phi(y)$ is the periodic function that describes the surface relief of the viscoelastic layer.

Depending on the load and sliding velocity, the contact region ω can be either simply connected or multiply connected. In the latter case, the contact region is a set of closed subregions ω_i arranged periodically along the y-axis. Let $-a(y)$ and $b(y)$ be the left and right ends of the contact region ω in the x-y plane. Since the functions $a(y)$ and $b(y)$ are periodic, they will be considered for a single period. At the right end $b(y)$ of the contact region, where the cylinder comes into contact with the viscoelastic layer, the layer displacement satisfies the condition:

$$u(b(y), y) = 0, \quad y \in [0, l]. \tag{5.52}$$

In virtue of smoothness of the contacting bodies, the contact pressure is zero at the entire boundary of the contact region, i.e.,

$$p(-a(y), y) = p(b(y), y) = 0, \quad y \in [0, l]. \tag{5.53}$$

It is assumed that $R \gg l$, where R is the cylinder radius and l is the period of the surface waviness. Hence, we can set that $k = R/l$ is an integer. Then the relations for the normal load P and moment M of sliding resistance acting upon the cylinder have the form:

$$\begin{aligned} P &= \int_0^R dy \int_{-a(y)}^{b(y)} p(x, y)dx = k \int_0^l dy \int_{-a(y)}^{b(y)} p(x, y)dx, \\ M &= \int_0^R dy \int_{-a(y)}^{b(y)} xp(x, y)dx = k \int_0^l dy \int_{-a(y)}^{b(y)} xp(x, y)dx. \end{aligned} \tag{5.54}$$

Relations (5.49)—(5.54) are the complete system of equations for the determination of the contact pressure distribution and analysis of the sliding friction force.

5.4.2 Method of solution

The problem is analyzed in the dimensionless form by using the following dimensionless parameters:

$$\overline{x}=\frac{x}{R},\quad \overline{y}=\frac{y}{R},\quad \overline{h}=\frac{h}{R},\quad \overline{l}=\frac{l}{R}=\frac{1}{k},$$

$$\overline{H}=\frac{H}{R},\quad \overline{D}=\frac{D}{R},\quad \alpha=\frac{T_\varepsilon}{T_\sigma},\quad \zeta=\frac{R}{T_\varepsilon V},$$

$$\overline{a}(\overline{y})=\frac{a(\overline{y}R)}{R},\quad \overline{b}(\overline{y})=\frac{b(\overline{y}R)}{R},\quad \overline{\Delta}(\overline{y})=\frac{\Delta(\overline{y}R)}{R},\quad \overline{\phi}(\overline{y})=\frac{\phi(\overline{y}R)}{R},$$

$$\overline{u}(\overline{x},\overline{y})=\frac{u(\overline{x}R,\overline{y}R)}{H},\quad \overline{p}(\overline{x},\overline{y})=\frac{p(\overline{x}R,\overline{y}R)}{E},$$

$$\overline{P}=\frac{P}{ER^2}=\int_0^1 d\overline{y}\int_{-\overline{a}(\overline{y})}^{\overline{b}(\overline{y})}\overline{p}(\overline{x},\overline{y})d\overline{x},$$

$$\overline{M}=\frac{M}{ER^3}=\int_0^1 d\overline{y}\int_{-\overline{a}(\overline{y})}^{\overline{b}(\overline{y})}\overline{x}\overline{p}(\overline{x},\overline{y})d\overline{x}. \tag{5.55}$$

In these coordinates, Eq. (5.49) has the form (from here and till the end of Section 5.4, bars over symbols are omitted):

$$u(x,y)-\frac{u'_x(x,y)}{\zeta}=p(x,y)-\frac{p'_x(x,y)}{\alpha\zeta}. \tag{5.56}$$

We consider this equation on the segment $y\in[0,1]$, which includes k periods of waviness of the surface relief. The function $u(x,y)$ satisfies the condition Eq. (5.52) at the right end $b(y)$ of the contact region, i.e.,

$$u(b(y),y)=0,\quad y\in[0,1]. \tag{5.57}$$

The function $p(x,y)$ satisfies the conditions Eq. (5.53), which can be rewritten in the dimensionless form as

$$p(-a(y),y)=p(b(y),y)=0,\quad y\in[0,1]. \tag{5.58}$$

Consider Eq. (5.56) as a differential equation for the unknown function $p(x, y)$ for any fixed $y \in [0, 1]$ and with the prescribed function $u(x, y)$ satisfying the condition Eq. (5.57). The boundary condition for this differential equation follows from Eq. (5.58): $p(b(y), y) = 0, \ y \in [0, 1]$. The solution to this equation has the form

$$p(x, y) = \alpha\zeta \int_{x}^{b(y)} \left(u(t, y) - \frac{u_t^{'}(t, y)}{\zeta} \right) e^{\alpha\zeta(x-t)} dt, y \in [0, 1]. \tag{5.59}$$

After integration by parts, Eq. (5.59) becomes

$$p(x, y) = \alpha u(x, y) - (\alpha - 1)\alpha\zeta \int_{x}^{b(y)} u(t, y) e^{\alpha\zeta(x-t)} dt, y \in [0, 1], \tag{5.60}$$

or

$$p(x, y) = u(x, y) - (\alpha - 1) \int_{x}^{b(y)} u_t^{'}(t, y) e^{\alpha\zeta(x-t)} dt, y \in [0, 1]. \tag{5.61}$$

The right end $b(y)$ of the contact region $[-a(y), b(y)]$ is determined from Eq. (5.57), whereas its left end, $-a(y)$, from the condition that follows from Eqs. (5.59) and (5.58):

$$\int_{-a(y)}^{b(y)} \left(u(t, y) - \frac{u_t^{'}(t, y)}{\zeta} \right) e^{\alpha\zeta(x-t)} dt = 0, y \in [0, 1]. \tag{5.62}$$

Eqs. (5.60) and (5.61) with the conditions Eqs. (5.57) and (5.62) completely define the contact pressure distribution for a prescribed dimensional function $u(x, y)$ (Eq. 5.50) or its dimensionless analog, which is introduced according to Eq. (5.55).

The dimensionless linear force P and moment M acting on the segment $y \in [0, 1]$ are calculated by the formulas

$$P = \iint_{\Omega} p(x, y)dxdy = k\iint_{\Omega_l} p(x, y)dxdy = \int_0^1 dy \int_{-a(y)}^{b(y)} p(x, y)dx$$

$$= k\int_0^{1/k} dy \int_{-a(y)}^{b(y)} p(x, y)dx,$$

$$M = \iint_{\Omega} xp(x, y)dxdy = k\iint_{\Omega_l} xp(x, y)dxdy = \int_0^1 dy \int_{-a(y)}^{b(y)} xp(x, y)dx$$

$$= k\int_0^{1/k} dy \int_{-a(y)}^{b(y)} xp(x, y)dx, \tag{5.63}$$

where $\Omega \equiv \{(x, y) : -a(y) \leq x \leq b(y), y \in [0, 1]\}$ is the contact region between the cylinder and the deformed surface of the viscoelastic layer for $y \in [0, 1]$, while $\Omega_l \equiv \{(x, y) : -a(y) \leq x \leq b(y), y \in [0, 1/k]\}$ is the contact region for a period, i.e., for $y \in [0, 1/k]$.

By using Eqs. (5.56), (5.57) and (5.59), the inner integrals in Eq. (5.63) can be expressed through the function $u(x, y)$ as

$$\int_{-a(y)}^{b(y)} p(x, y)dx = \int_{-a(y)}^{b(y)} u(x, y)dx + \frac{u(-a(y), y)}{\zeta},$$

$$\int_{-a(y)}^{b(y)} xp(x, y)dx = \int_{-a(y)}^{b(y)} \left(x + \frac{\alpha - 1}{\alpha\zeta}\right)u(x, y)dx - \frac{u(-a(y), y)}{\zeta}\left(a(y) + \frac{1}{\alpha\zeta}\right),$$

$$y \in [0, 1]. \tag{5.64}$$

Note that the function $u(x, y)$ prescribes the shape of the deformed surface of the viscoelastic layer inside the contact region.

5.4.3 Calculation of the contact characteristics

Since the strains are assumed to be small, we can approximate the profile of the contacting part of the cylinder by the function $f(x) = \frac{x^2}{2R}$, where R is the cylinder radius. Then the expression Eq. (5.50) for the function $u(x, y)$ in the dimensionless form Eq. (5.55), with taking into account Eq. (5.57), takes the form:

$$u(x, y) = \frac{1}{H}\left(\Delta(y) - \frac{x^2}{2}\right) = \frac{b^2(y) - x^2}{2H}, \quad (5.65)$$

where

$$b(y) = \sqrt{2\Delta(y)}. \quad (5.66)$$

The dimensionless function $\Delta(y)$ can be calculated by the following formula (see Eqs. 5.51 and 5.55):

$$\Delta(y) = D - \phi(y), \quad (5.67)$$

where D is the dimensionless indentation of the cylinder into the viscoelastic layer and $\phi(y)$ is the dimensionless periodic function that describes the surface relief of the viscoelastic layer. It is assumed that if the value of the function $\Delta(y)$ calculated by Eq. (5.67) is negative, then for this y the cylinder and the layer are not in contact. Note that the function $u(x, y)$ given by Eq. (5.65) under the condition Eq. (5.66) satisfies the condition Eq. (5.57).

For the function $u(x, y)$ given by Eq. (5.65) the unknown contact pressure $p(x, y)$ follows from (5.59), (5.60), or (5.61) and can be calculated as

$$p(x, y) = \frac{1}{H}\left\{\frac{b^2(y) - x^2}{2} + \frac{\alpha - 1}{\alpha\zeta}x + \frac{\alpha - 1}{(\alpha\zeta)^2}\left[1 - (\alpha\zeta b(y) + 1)e^{\alpha\zeta(x - b(y).)}\right]\right\},$$

$$-a(y) \leq x \leq b(y), \quad y \in [0, 1], \quad (5.68)$$

After calculating the inner integrals in Eq. (5.64) with the function $u(x, y)$ given by Eq. (5.65), we obtain the following expressions:

$$\int_{-a(y)}^{b(y)} p(x, y)dx = \frac{1}{2H}\left[\frac{(b(y)+a(y))^2(2b(y)-a(y))}{3} + \frac{b^2(y)-a^2(y)}{\zeta}\right],$$

$$\int_{-a(y)}^{b(y)} xp(x, y)dx = \frac{1}{2H}\left[\left(\frac{b^2(y)-a^2(y)}{2}\right)^2 + \frac{\alpha-1}{\alpha\zeta}\cdot\frac{(b(y)+a(y))^2(2b(y)-a(y))}{3}\right]$$

$$-\frac{b^2(y)-a^2(y)}{2H\zeta}\left(a(y)+\frac{1}{\alpha\zeta}\right), \quad y\in[0, 1], \tag{5.69}$$

which allow us to calculate the dimensionless linear force P and moment M by formulas (5.63).

In Eqs. (5.68) and (5.69), the right end of the contact region $[-a(y), b(y)]$ is determined by relation Eq. (5.66), whereas its left end, as it follows from Eqs. (5.58), (5.62), and (5.68), is a solution of the transcendental equation

$$\frac{b^2(y)-a^2(y)}{2} - \frac{\alpha-1}{\alpha\zeta}a(y) + \frac{\alpha-1}{(\alpha\zeta)^2}\left[1-(\alpha\zeta b(y)+1)e^{-\alpha\zeta(a(y)+b(y))}\right] = 0. \tag{5.70}$$

It is shown (Goryacheva and Goryachev, 2020) that this equation has a unique solution.

5.4.4 Analysis of the calculation results

Relations (5.57), (5.62), (5.63), (5.68), and (5.69) were used for the calculation of the contact pressure distribution and integral characteristics of contact (the moment and indentation of the cylinder into the viscoelastic foundation), provided that the linear force applied to the cylinder and its sliding velocity are known. The surface relief is described by the function $\phi(y) = h\sin^2 \pi y$.

The analysis of the obtained relations for the contact pressure, linear force, and moment shows that the contact characteristics depend on six dimensionless parameters: α, H, P, ζ, k, and h (Eq. 5.55). We assume that the relative thickness of the viscoelastic layer and linear load applied to the cylinder are constant ($H = 100$, $P = 0.001$).

In Fig. 5.12, plots of the contact pressure distribution are presented for the cross-sections cutting through the axis of the cylinder in parallel ($x = 0$) and perpendicular ($y = 0$ and $y = l/2$) to its generatrix for various values of the parameter ζ, which is inversely proportional to the sliding velocity of the cylinder. The calculations were carried out for the prescribed parameters of the surface relief ($k = 10$ $h/l = 1$) and for $\alpha = 5$.

The results of calculation indicate that as the sliding velocity of the cylinder increases and, accordingly, the parameter ζ decreases, the contact pressures grow, and the contact becomes incomplete. The pressure distribution in the cross-sections $y =$ const is asymmetric, the asymmetry being more significant for ζ close to 1.

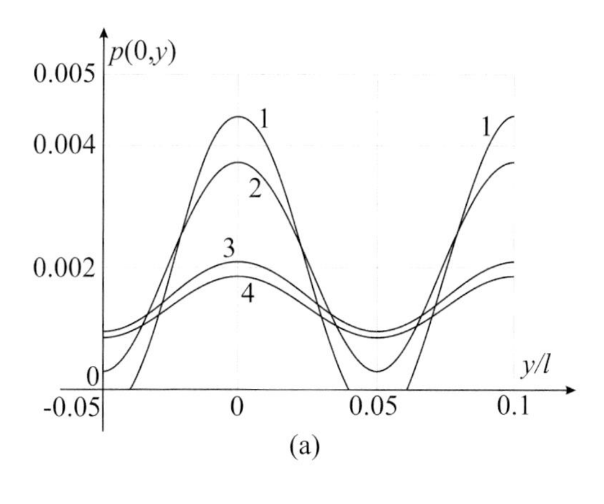

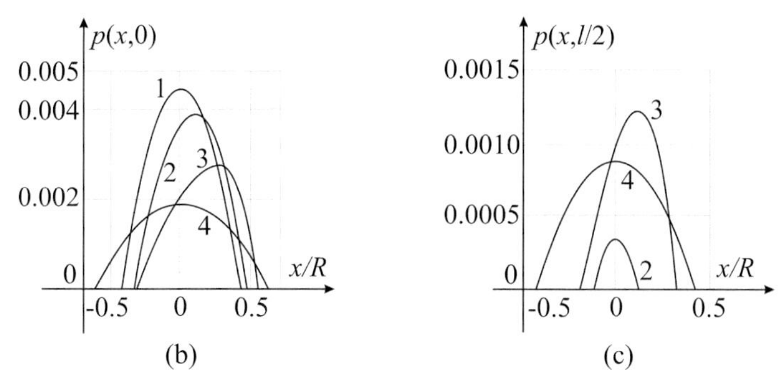

Figure 5.12 Contact pressure distribution for $x = 0$ (a), $y = 0$ (b), и $y = l/2$ (c), and for $k = 10$, $h/l = 1$, $\alpha = 5$ and the following values of the parameter ζ: $\zeta = 10^{-4}$ (curves 1), $\zeta = 10^{-0,5}$ (curves 2), $\zeta = 10^{0,2}$ (curves 3), $\zeta = 10^{4}$ (curves 4).

The contact region between the cylinder and the wavy foundation can be both continuous and consisting of separate spots. The results presented in Fig. 5.12 show that for $\zeta = 10^{-4}$, the contact region is not continuous. The shape and size of the contact region depend on the mechanical and geometric parameters of the foundation and also on the sliding velocity. In Fig. 5.13, the contact regions are presented for various values of the parameter ζ characterizing the sliding velocity (Fig. 5.13a) and the parameter h/l characterizing the relief waviness amplitude (Fig. 5.13b).

As the parameter ζ increases (the sliding velocity decreases), the contact passes from discrete to continuous, for which the contact region in the y-direction is bounded by wavy lines. The contact region is asymmetric: it is shifted in the direction of sliding of the cylinder (in the x-direction). As the amplitude of waviness of the foundation increases (parameter $h/\ l$ increases), the amplitude of oscillations of the width of the contact region along the y-axis also increases. For a foundation with flat surface, the contact region is a strip of constant width shifted in the x-direction (curves 4 in Fig. 5.13b).

In Fig. 5.14, the integral characteristics of contact are presented: the moment of resistance M/P and the indentation D of the cylinder that are calculated for various characteristics of surface relief of the foundation.

From the results obtained (Fig. 5.14a), it follows that the moment of resistance to sliding of the cylinder over the viscoelastic foundation is a nonmonotonic function of the sliding velocity. As the parameter k defined by Eq. (5.69) decreases, the maximum value of the moment increases and shifts to higher velocities (smaller values of ζ). The parameter $k = \ R/\ l$

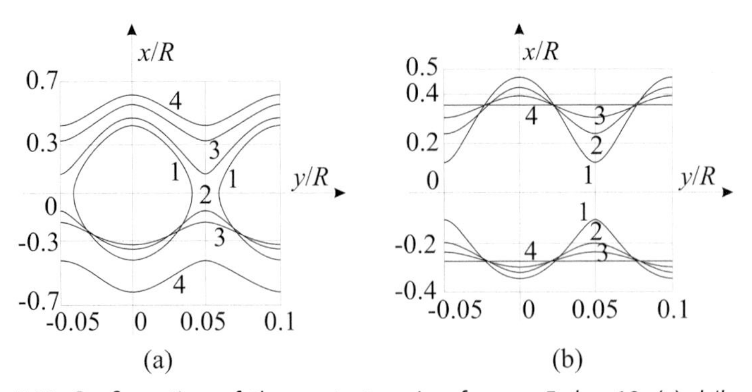

Figure 5.13 Configuration of the contact region for $\alpha = 5$, $k = 10$; (a): $h/l = 1$ and $\zeta = 10^{-4}$ (curve 1), $\zeta = 10^{-0.5}$ (curve 2), $\zeta = 10^{0.2}$ (curve 3), $\zeta = 10^{4}$ (curve 4); (b): $\zeta = 10^{-0.5}$ and $h/l = 1$ (curve 1), $h/l = 0{,}6$ (curve 2), $h/l = 0{,}3$ (curve 3), $h/l = 0$ (curve 4).

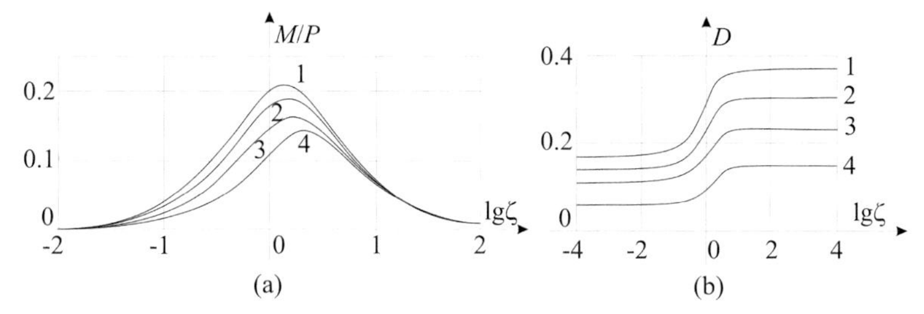

Figure 5.14 Moment of resistance (a) and indentation of the cylinder (b) versus the parameter ζ for $\alpha = 5$, $h/l = 1$ and various values of the parameter k: $k = 1$ (curve 1), $k = 2$ (curve 2), $k = 5$ (curve 3), $k = 100$ (curve 4).

characterizes the scale of roughness in comparison to the radius of the cylinder. The dependence of the indentation of the cylinder into the viscoelastic foundation on the parameter ζ has three characteristic parts: for small ($\zeta < 10^{-1}$) and large ($\zeta > 10$) values of this parameter, the indentation is close to constant, while for intermediate values the indentation grows as the parameter ζ increases. As the parameter k decreases, the indentation of the cylinder into the wavy viscoelastic foundation and the moment of resistance to sliding increase.

The results presented in Fig. 5.15 allow us to analyze the combined influence of the viscosity of the layer and its microrelief on the moment of resistance to sliding of the cylinder. An increase of the waviness amplitude leads to an increase in the moment of resistance, provided that the viscosity of the foundation and the period of waviness remain the same (see Fig. 5.15a). An increase in the viscosity leads to an increase in the moment

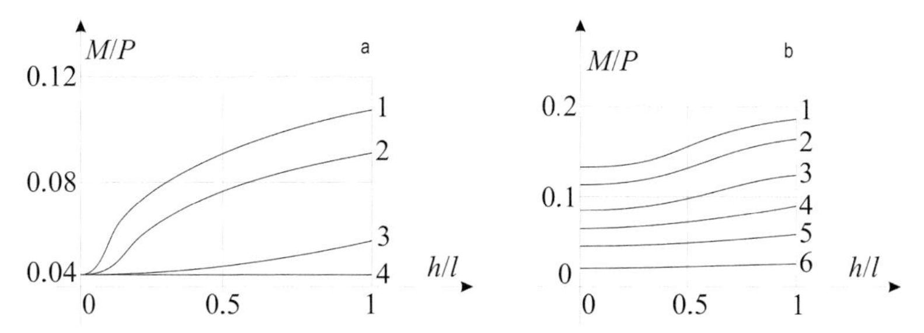

Figure 5.15 Moment of resistance as a function of the parameter h/l for $\alpha = 5$, $\zeta = 10^{-0.5}$, $P = 0.01$ and various values of the parameter R/l (a): $R/l = 1$ (curve 1), $R/l = 2$ (curve 2), $R/l = 10$ (curve 3), $R/l = 1000$ (curve 4); and also for various values of the parameter α (b): $\alpha = 100$ (curve 1), $\alpha = 50$ (curve 2), $\alpha = 20$ (curve 3), $\alpha = 10$ (curve 4), $\alpha = 5$ (curve 5), $\alpha = 2$ (curve 6).

of resistance (Fig. 5.15b). The dependence of the moment of resistance on the viscosity is close to exponential for small relative heights of asperities (for $h/l < 0.5$). For more viscous materials (Fig. 5.15b, curves 1—3), an increase in the relative height of asperities leads to a considerable growth of the moment of resistance to sliding of the cylinder.

So, the combined effects of the layer surface microgeometry and its viscosity on the friction force at an arbitrary fixed sliding velocity has been analyzed in this part. The calculation results can be used for controlling the friction force in sliding over the viscoelastic base by choosing the parameters of its microgeometry for given loading-velocity characteristics and mechanical properties of the base.

5.5 Sliding contact of a punch with periodic microgeometry over the viscoelastic foundation (3-D analysis)

In this section, a model of surface microgeometry described by a doubly periodic function is used to study the combined effect of surface roughness and imperfect elasticity on the contact characteristics and the friction force. The sliding contact of a punch having regular surface microrelief and a viscoelastic layer characterized by the spectrum of relaxation times are considered.

5.5.1 Contact problem formulation

In Fig. 5.16, the scheme of contact is presented for a punch with regular surface relief (1) that slides in the x-direction with the velocity V over a

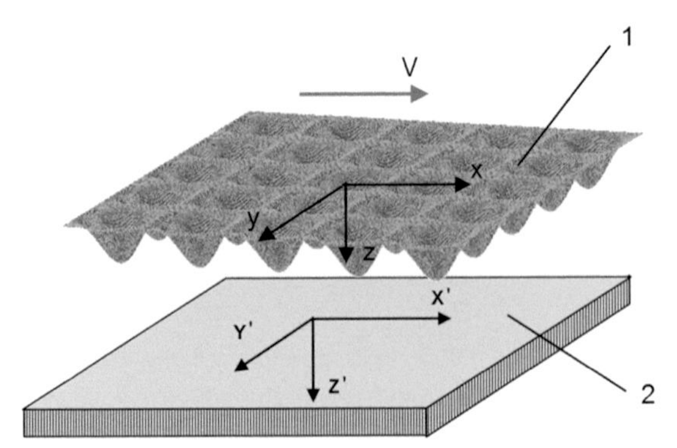

Figure 5.16 Scheme of contact between the punch with regular relief (1) and the viscoelastic layer (2) lying on the rigid base (3).

viscoelastic foundation (2). The surface relief of the punch is described by a doubly periodic function $f(x, y)$: $f(x + nl_1, y + ml_2) = f(x,y)$; $n = 1,2,\ldots\infty$; $m = 1,2, \ldots, \infty$; where l_1 and l_2 are the distances between neighboring asperities of the punch in the x- and y-directions, respectively. We introduce the system of coordinates (x', y', z') fixed on the foundation and the system of coordinates (x, y, z) moving together with the punch. These systems of coordinates are related as

$$x' = x + Vt, \quad y' = y, \quad z' = z. \tag{5.71}$$

In the moving system of coordinates (x, y, z), displacements and stresses do not depend implicitly on time t and are functions of the coordinates (x, y, z).

The contact region Ω between the punch and the foundation is a set of discrete contact spots, for which the normal displacement of the boundary of the viscoelastic foundation $u(x, y)$ satisfies the contact condition $(z = 0)$:

$$u(x, y) = D + f(x, y), \quad (x, y) \in \Omega, \tag{5.72}$$

where D is the indentation of the punch into the viscoelastic foundation.

The contact pressure $p(x, y)$ satisfies the periodicity condition

$$p(x, y) = p(x + nl_1, y + ml_2); m = 1, 2, \ldots, \infty; \; n = 1, 2, \ldots, \infty, \tag{5.73}$$

and it also satisfies the condition

$$p(x, y) = 0, (x, y) \notin \Omega. \tag{5.74}$$

At the boundary Γ_Ω of the region Ω, we have the condition of continuity

$$p(x, y)|_{\Gamma_\Omega} = 0. \tag{5.75}$$

5.5.2 Material model

As a model of the viscoelastic foundation, we consider a viscoelastic layer of thickness H bonded to the rigid base. The layer properties are described by the 1-D Kelvin model with a spectrum of relaxation times, in accordance with which the strain of the layer is related to the pressure as follows:

$$\varepsilon(t) = \frac{1}{E} p(t) + \frac{1}{E} \int_{-\infty}^{t} p(\tau) \sum_{i=1}^{n} k_i \exp[-(t-\tau)\beta_i] d\tau. \tag{5.76}$$

Taking into account the relation for the strain $\varepsilon = u(x,y)/H$, we can obtain the following relation for the displacement $u(x,y)$ of the upper boundary of the layer and the contact pressure $p(x,y)$ in the moving system of coordinates:

$$u(x,y) = \frac{H}{E}p(x,y) + \frac{H}{EV}\int_{\infty}^{x} p(\xi,y)\sum_{i=1}^{n} k_i \exp\left[-\left(\frac{\xi - x}{V}\right)\beta_i\right]d\xi. \quad (5.77)$$

5.5.3 Method of solution

To solve the problem, we consider a cell of periodicity of the punch relief (Fig. 5.17). This cell is divided into m thin strips parallel to the sliding direction. For each strip, a 2-D contact problem is to be solved. This method is exact for the case of 1-D viscoelastic foundation.

The function of pressure $p(x)$ in the interval $x \in (-l_1/2, l_1/2)$ is represented as a piecewise constant function: $p(x) = p_i$ for $x \in (x_i, x_{i+1})$, $i = 1, \ldots K$, where $x_i = -l_1/2 + (i-1)l_1/(K-1)$ (Fig. 5.18). Here, K is the number of intervals.

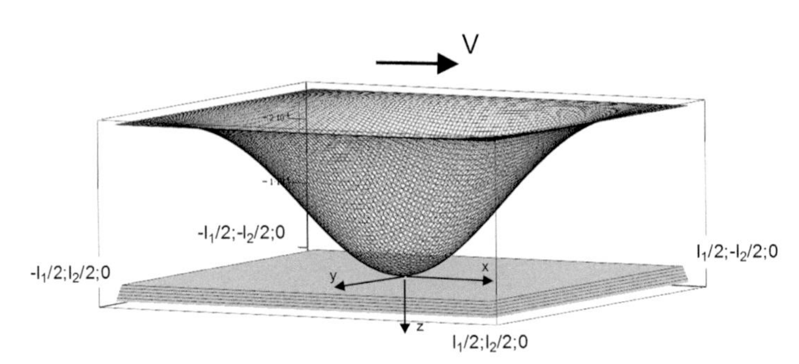

Figure 5.17 Scheme of contact in a cell of periodicity.

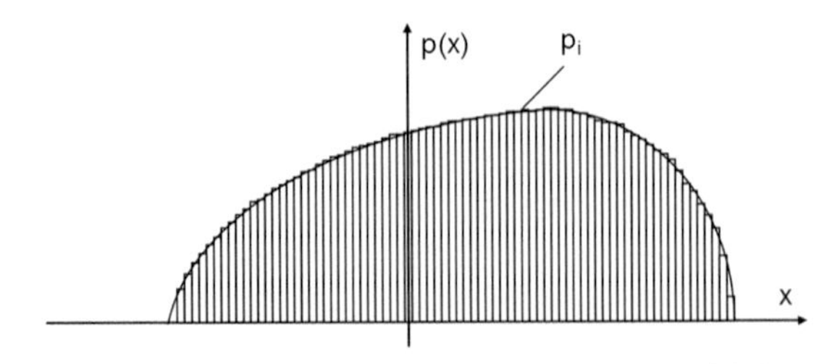

Figure 5.18 A piecewise constant function of pressure $p(x)$ for a j-th strip.

The normal displacements of the boundary of the viscoelastic layer are then represented as $u(x) = u_{i'}$, $x \in (x_{i'}, x_{i'+1})$, $i' = 1, \ldots K$.

The relation between the normal displacement of the boundary and the normal pressure acting on this boundary has the form:

$$\{u\} = [S]\{p\}, \tag{5.78}$$

where $[S]$ is a matrix with the following elements $S_{i'j'}$

$$S_{i'j'} = \frac{H}{E} + \frac{H}{E}\sum_{i=1}^{n}\frac{k_i}{\beta_i}\left(\exp\left[-\frac{(j'-i')\beta_i}{Kf}\right] - \exp\left[-\frac{(j'-i'+1)\beta_i}{Kf}\right]\right)_{\text{for } j' \geq i'}$$

$$+\frac{H}{E}\sum_{i=1}^{n}\sum_{j=1}^{M}\frac{k_i}{\beta_i}\left(\exp\left[-\frac{(jK+j'-i')\beta_i}{Kf}\right]\right.$$

$$\left.-\exp\left[-\frac{(jK+j'-i'+1)\beta_i}{Kf}\right]\right). \tag{5.79}$$

Here, $f = \frac{V}{l}$ is the frequency of interaction. The material is described by a set of matrices $[S]^k$ for each value of the frequency f_k.

The piecewise constant function p_i satisfies the condition Eq. (5.74) of zero contact pressure outside the contact region. The displacement u_i satisfies the contact condition Eq. (5.72).

Eq. (5.78) and condition Eq. (5.72) together with (5.73) and (5.74) form a system of linear equations for the determination of the contact pressure.

For a given indentation D, the contact area is calculated by the iteration method. As a convergence conditions, we use the condition of absence of intervals with negative contact pressure and the condition of zero pressure at the boundary of the contact area (Eq. 5.75).

The projections of the forces acting on an asperity are calculated by the integration of the x- and z-projections of the contact pressure over an elementary contact spot ω. For the case $l_1 = l_2 = l$, they can be calculated as

$$T = \iint_{\omega} p(x,y)\sin\alpha\, dxdy = l^2\sum_{i=1}^{m}\sum_{j=1}^{m} p_{i,j}\sin a_i$$

$$P = \iint_{\omega} p(x,y)\cos\alpha\, dxdy = l^2\sum_{i=1}^{m}\sum_{j=1}^{m} p_{i,j}\cos a_i, \tag{5.80}$$

where α is the angle between the normal to the contact surface and the z-axis. The normal component P is equal to the normal load applied to the punch. The tangential component T is directed opposite to the direction of sliding of the punch and is a deformation component of the friction force associated with hysteretic losses in the viscoelastic material.

The deformation component of the coefficient of friction can be calculated as

$$\mu = \frac{T}{P} = \frac{\sum_{i=1}^{m}\sum_{j=1}^{m} p_{i,j} \sin a_i}{\sum_{i=1}^{m}\sum_{j=1}^{m} p_{i,j} \cos a_i}. \tag{5.81}$$

5.5.4 Analysis of the results of calculation

We consider a viscoelastic material, for which the effective compliance E_{1i} versus relaxation time $T_{\sigma i}$ is described by the graph presented in Fig. 5.19.

Based on the dependence presented in Fig. 5.19, we calculate the values of the relaxation spectrum E, k_i, and β_i, which are used in Eq. (5.77), by using the following relations:

$$k_i = \frac{1000 E_{1i}}{E\lambda_i}, \beta_i = \frac{1001}{\lambda_i}, E = \sum_{1}^{n} E_i. \tag{5.82}$$

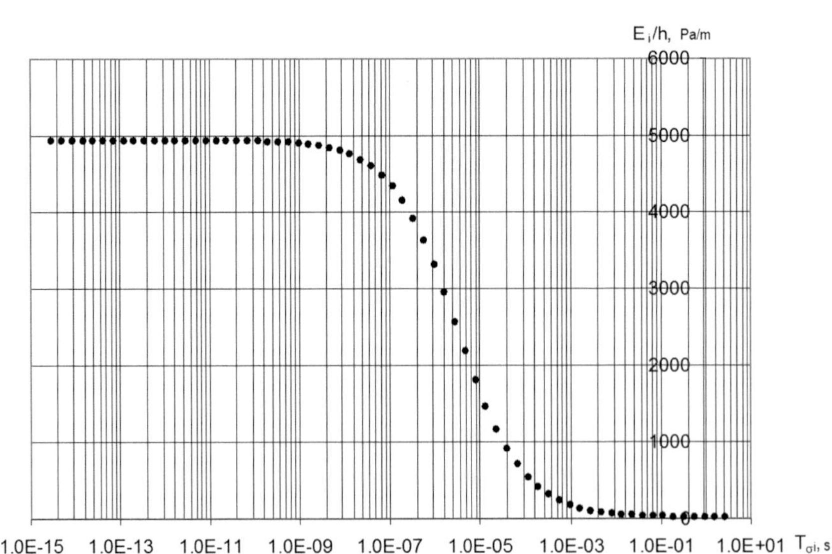

Figure 5.19 Effective compliance (E_{1i}/H) as a function of relaxation time ($T_{\sigma i}$) for the viscoelastic material under consideration.

Let us calculate the contact pressure distribution for two regular surfaces that are characterized by the same period and height of asperities ($h = 0.0002$ m and $l_1 = l_2 = 0.001$ m) and different shape of the asperities (Fig. 5.20). In both cases, the normal load acting on an asperity is $P = 8 \cdot 10^{-6}$ N and the sliding velocity is $V = 0.5$ m/s. Since the load P and the nominal pressure p_0 are related by the expression $P = p_0 l^2$, we can calculate the nominal pressure $p_0 = 8$ Pa. The first regular surface under consideration has spherical asperities, while the second one is described by a doubly periodic sinusoidal function:

$$f(x, y) = h - \frac{h}{4}\left(\cos\left(\frac{2\pi x}{l}\right) + 1\right)\left(\cos\left(\frac{2\pi y}{l}\right) + 1\right), \quad (5.83)$$

where h and l are the height of asperities and distance between them, respectively.

The contact pressure distributions for a half of the cell of periodicity are presented in Fig. 5.21.

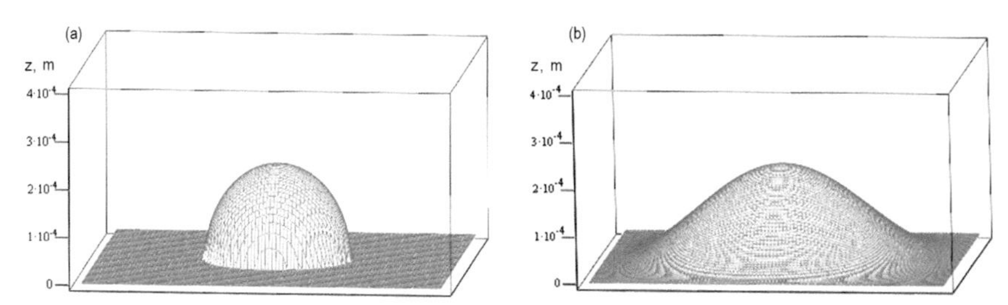

Figure 5.20 Asperities of different shape: (a) spherical and (b) described by the function Eq. (5.83).

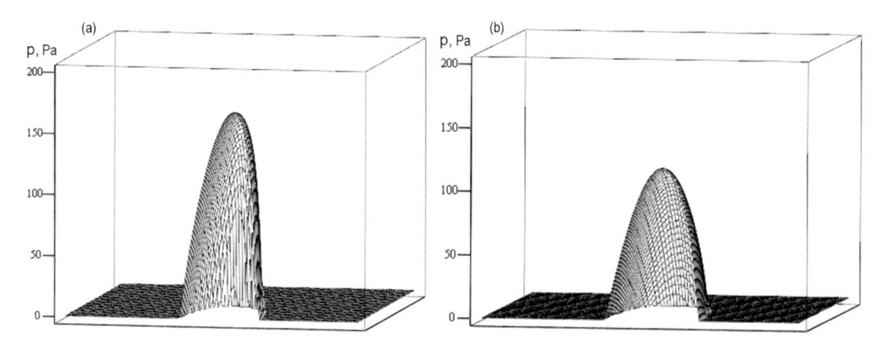

Figure 5.21 Contact pressure distribution on a half of a periodicity cell for the regular surface with spherical asperities (a) and described by the function Eq. (5.83) (b).

The comparison of the results presented in Fig. 5.21 allows us to conclude that for a spherical asperity (Fig. 5.21a), the maximum contact pressure is higher and the contact area is smaller than those for an asperity of the sinusoidal profile (Fig. 5.21b). In Fig. 5.22, the contact pressure distributions in the central strip and the corresponding contours of the contact area are presented for $V = 0.5$ m/s and $P = 8 \cdot 10^{-6}$ N. It is characteristic that the contact area is shifted along the x-axis in the direction of the punch motion, and the pressure distribution is asymmetric with respect to the y-z plane.

Effect of the sliding velocity on the contact characteristics is illustrated in Fig. 5.23. The results are calculated for a regular indenter of sinusoidal shape

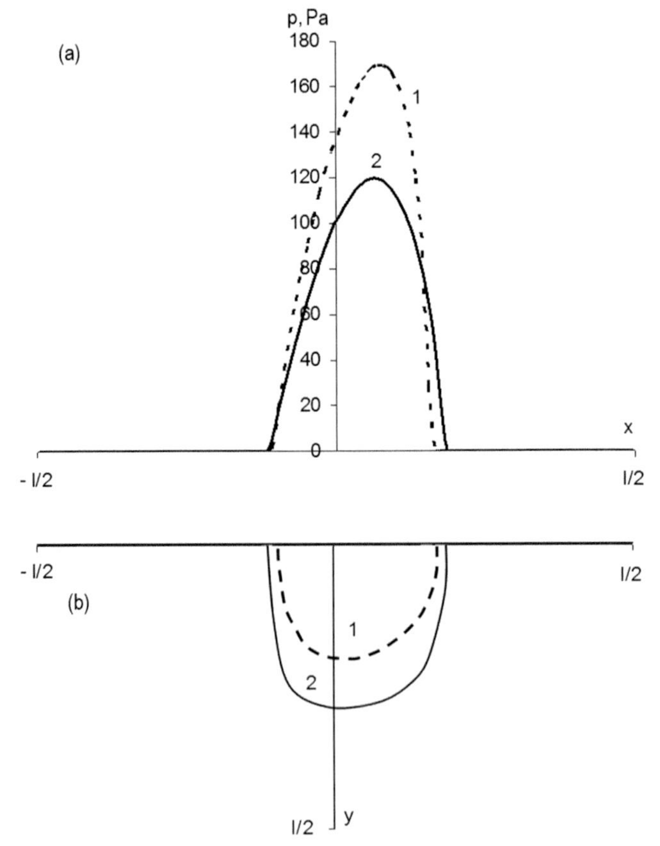

Figure 5.22 Contact pressure distribution in the central strip (a) and contour of the half contact area (b) for the regular surface with spherical asperities (curves 1) and described by the function Eq. (5.83) (curves 2).

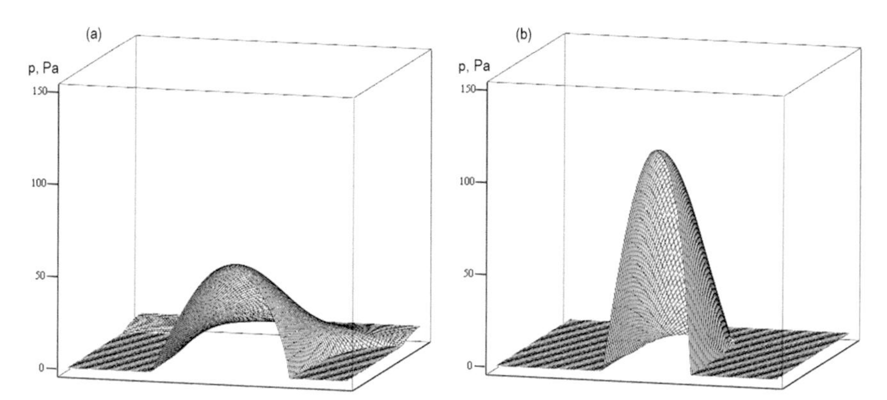

Figure 5.23 Contact pressure distribution for the load $P_z = 1.16 \cdot 10^{-5}$ N and different sliding velocities $V = 0.1$ m/s (a) and $V = 2$ m/s (b).

described by Eq. (5.83) with the asperity height $\mathrm{H} = 0.0001$ m and period $l = 0.001$ m. The load applied to an asperity is $P = 1.16 \cdot 10^{-5}$ N, which corresponds to the nominal pressure $p_0 = 11.6$ Pa. In Fig. 5.23, the contact pressure distribution is shown for two sliding velocities: 0.1 m/s and 2 m/s. For the velocity 0.1 m/s, the contact areas merge in the direction orthogonal to the sliding direction. As the sliding velocity increases ($V = 2$ m/s), the contact pressure becomes asymmetric, and the contact area decreases (floating effect). With further increase in velocity, the contact area continues to decrease, whereas the pressure distribution becomes close to symmetric.

The asymmetry in the contact pressure distribution and contact region configuration lead to a tangential force T acting on an asperity in the direction opposite to the sliding direction. This force is the hysteretic (deformation) friction force, and it can be calculated by the first relation of Eq. (5.80). In Fig. 5.24, the hysteretic component of the coefficient of friction calculated by Eq. (5.81) is presented as a function of the sliding velocity. This function is nonmonotone, its maximum being higher for the regular surface with closer arranged asperities.

5.6 Combined effect of hysteresis and adhesion in sliding of a wavy surface

In this section, the additional (to the imperfect elasticity and microgeometry) effect accounted for by the adhesive attraction between the contacting surfaces on the contact characteristics is modeled and analyzed.

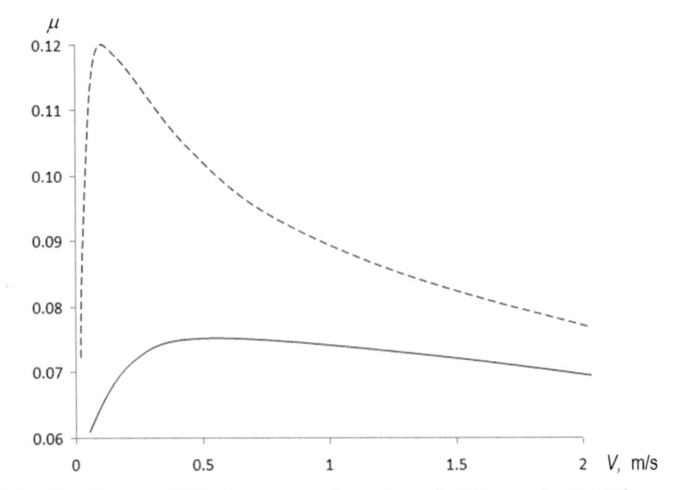

Figure 5.24 Coefficient of friction μ as a function of sliding velocity V for two regular sinusoidal surfaces with the same height of asperities ($h = 0.0001$ m) and different distance between them (solid line corresponds to the distance 0.001 m, dashed line to 0.0005 m) under the constant nominal pressure $p_0 = 11.6$ Pa.

5.6.1 Periodic contact problem formulation

Consider a rigid wavy surface sliding with the velocity V along the x-axis on the viscoelastic foundation (Fig. 5.16). The shape of the wavy surface is described by the periodic function given by Eq. (5.83).

Mechanical properties of the viscoelastic layer are modeled by the Kelvin model given by Eq. (5.49).

Let the system of coordinates (x', y', z') be connected with the viscoelastic foundation, and the system of coordinates (x, y, z) with the sliding wavy surface so that relation Eq. (5.71) is satisfied. In the moving system of coordinates (x, y, z),Eq. (5.98) has the form

$$u - VT_\varepsilon \frac{\partial u}{\partial x} = \frac{H}{E}\left(p - T_\sigma V \frac{\partial p}{\partial x}\right). \quad (5.84)$$

To take into account adhesive (molecular) attraction between the surfaces, introduce a negative adhesive stress $p = -p_a(\delta)$ acting on the boundary of the viscoelastic foundation, where δ is the value of gap between the surfaces. We use the Maugis model in which the dependence of the adhesive stress on the gap between the surfaces has a form of one-step function (see Section 2.1.1):

$$p_a(z) = \begin{cases} p_0, & 0 < z \le h_0 \\ 0, & z > h_0 \end{cases}, \quad (5.85)$$

where δ_0 is the maximum value of gap for which the adhesive attraction acts. The specific work of adhesion w_a is specified by the relation:

$$w_a = \int_0^{+\infty} p_a(z)dz = p_0 h_0. \tag{5.86}$$

In the moving system of coordinates *(x, y, z),* which is fixed at the indenter surface, the following boundary conditions for stresses and displacements are satisfied at the half-space surface ($z = 0$):

$$\begin{array}{ll} u(x, y) = f(x, y) + D, & (x, y) \in \Omega^c \\ p(x, y) = -p_0, & (x, y) \in \Omega^a \\ p(x, y) = 0, & (x, y) \notin \Omega^c \cup \Omega^a \end{array} . \tag{5.87}$$

Here Ω^c is the contact region, Ω^a is the region in which the adhesive stress $-p_0$ acts, and D is penetration of an asperity into the foundation. The equilibrium condition is also satisfied in the periodicity cell:

$$P = \iint_{\Omega^c \cup \Omega^a} p(x, y)dxdy. \tag{5.88}$$

5.6.2 Solution by using the strip method

The technique of solution of the sliding contact problem for a viscoelastic foundation and a wavy surface taking into account adhesion was developed previously for a 2-D surface (Goryacheva and Makhovskaya, 2010) and for a 3-D surface by using the strip method (Goryacheva and Makhovskaya, 2015). Application of the strip method is similar to that presented in Section 5.5.3. The difference is that in the present case, the contact problem solution for each strip is obtained in the closed form.

The square region $x \in (-l/2; l/2)$; $\quad y \in (-l/2; l/2)$ is divided into $2N$ strips of equal thickness Δ that are parallel to direction of sliding (axis Ox) (Fig. 5.25). The normal displacement of the center of a strip j is

$$D_j = \frac{h}{2}\left(\cos\left(\frac{2\pi y_j}{l}\right) - 1\right) + D, \tag{5.89}$$

where $y_j = j\Delta$. The penetration D_j is maximal in the central strip $j = 0$. The shape of the indenter in the j-th strip is

$$f\left(x, y_j\right) = \frac{h_j}{2}\left(\cos\frac{2\pi x}{l} - 1\right), h_j = \frac{h}{2}\left(\cos\frac{2\pi y_j}{l} + 1\right). \tag{5.90}$$

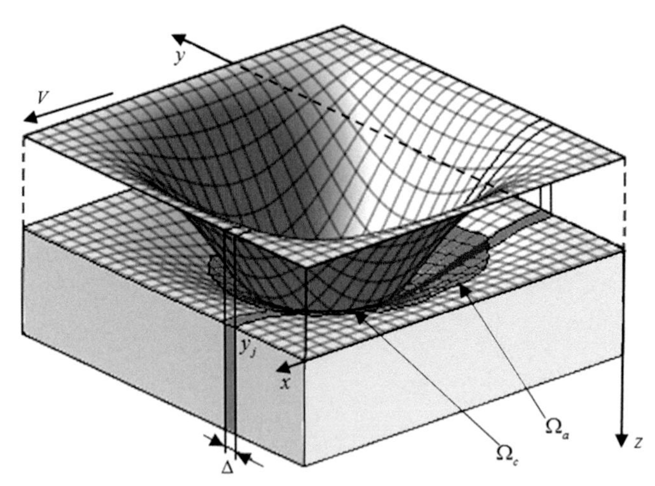

Figure 5.25 Scheme of calculation by the strip method in a periodicity cell.

In each strip, the contact problem is formulated and solved independently. For a j-th strip, the conditions for the displacement $u(x, y_j)$ and pressure $p(x, y_j)$ follow from Eq. (5.87):

$$\begin{aligned} w(x, y_j) &= f(x, y_j) + d_j, && x \in \Omega_j^c \\ p(x, y_j) &= -p_0, && x \in \Omega_j^a \\ p(x, y_j) &= 0, && x \notin \Omega_j^c \cup \Omega_j^a \end{aligned}, \quad (5.91)$$

where Ω_j^c is the contact region in the j-th strip, and Ω_j^a is the adhesion region in the j-th strip.

Thus, the original 3-D contact problem is reduced to a 2-D contact problem to be solved in each strip to determine the unknown contact pressure $p(x, y_j)$ in the contact region $x \in \Omega_j^c$ and the displacement $u(x, y_j)$ of the boundary of the viscoelastic foundation in the adhesion region $x \in \Omega^a$. To more precisely write the boundary conditions for the contact pressures and displacements in a strip (in particular, to determine the configuration of the contact and adhesion regions Ω_j^c and Ω_j^a), one should take into account a regime of interaction, which is specific for each strip.

The following three regimes of contact in a strip are considered (Fig. 5.26): saturated contact (a), discrete contact with saturated adhesion (b), and discrete contact with discrete adhesion (c). In a j-th strip, one of these regimes occurs, depending on the penetration D_j of the indenter in this strip. Below, we present the algorithm for the determination of a contact regime and contact problem solution in a j-th strip.

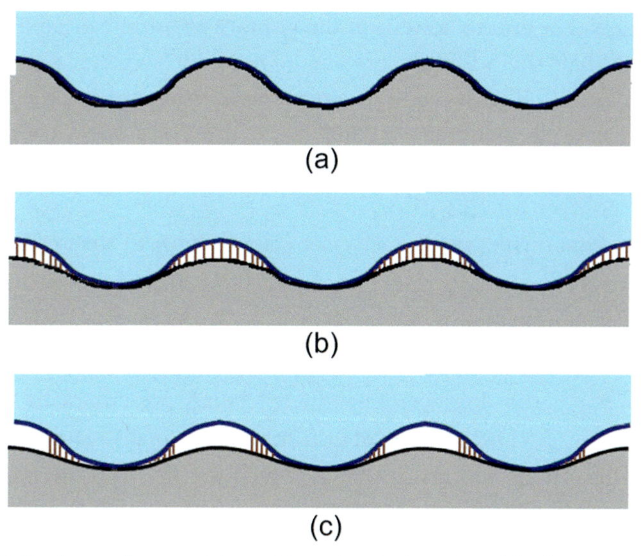

Figure 5.26 Regimes of contact in a strip: saturated contact (a), discrete contact with saturated adhesion (b), and discrete contact with discrete adhesion (c).

Saturated contact (Fig. 5.26a). First, the possibility of saturated contact is considered. In this case, the surfaces of the viscoelastic foundation and rigid indenter are in contact over the entire length of the strip. The displacement $u(x, y_j)$ satisfies the first of the conditions Eq. (5.91) in the whole interval $x \in [-l/2, l/2]$. We substitute this condition into Eq. (5.84) to obtain an ordinary linear differential equation for the pressure $p(x, y_j)$. By solving this equation in the interval $x \in [-l/2, l/2]$ with the condition of periodicity $p(x - l/2, y_j) = p_j(x + l/2, y_j)$ as the boundary condition, we obtain

$$p_j(x) = \frac{E}{2H\left(l^2 + 4\pi^2 T_\sigma^2 V^2\right)} \left[h_j(l^2 + 4\pi^2 T_\varepsilon T_\sigma V^2)\cos\frac{2\pi x}{l} + 2\pi l h_j V(T_\varepsilon - T_\sigma)\sin\frac{2\pi x}{l} + \left(4\pi^2 T_\sigma^2 V^2 + l^2\right)\left(2D_j - h_j\right) \right]. \tag{5.92}$$

The contact in the j-th strip is saturated provided that the condition for the minimum contact stress is satisfied:

$$\min(p(x, y_j)) \geq -p_0, \tag{5.93}$$

which means that due to adhesion, the contact pressure can be negative but not smaller than the adhesion pressure $-p_0$. If the contact pressure calculated by Eq. (5.92) in a j-th strip does not satisfy the condition Eq. (5.93), then the penetration D_j of this strip is insufficient for the contact to be saturated. In this case, it is necessary to consider the next possible regime of contact for this strip.

Discrete contact with saturated adhesion (Fig. 5.26b). In this regime, contact occurs over the region $\Omega^c{}_j = \left(-a_j, b_j\right)$ and the adhesion pressure $-p_0$ acts over the remaining regions $\Omega_j^a = \left(-l/2, -a_j\right) \cup \left(b_j, l/2\right)$. The differential Eq. (5.99) is solved in the contact interval $-a_j < x < b_j$ for the contact pressure $p(x, y_j)$, the displacement $u(x, y_j)$ being prescribed by the first of conditions Eq. (5.91). In the adhesion intervals $-l/2 < x < -a_j$ and $b_j < x < l/2$, the differential Eq. (5.84) is solved for the displacement $u(x, y_j)$, the pressure $p(x, y_j)$ being prescribed by the second condition of Eq. (5.91). For these two equations to be solved, we need two corresponding boundary conditions. Apart from this, we need two conditions for the determination of the unknown end points a_j and b_j of the contact interval. As such conditions, we use the conditions of continuity of the functions $p(x, y_j)$ and $u(x, y_j)$ at the points $x = -a_j$ and $x = b_j$ and the conditions of periodicity for these two functions. As a result, we get two nonlinear equations for the numerical calculation of the values a_j and b_j. The pressure $p(x, y_j)$ in the contact interval $-a_j < x < b_j$ is then determined as

$$
\begin{aligned}
p_j(x) = & -p_0 e^{(x-a_j)/T_\sigma V} + \frac{Eh_j}{2H(l^2 + 4\pi^2 T_\sigma{}^2 V^2)} \left[(l^2 + 4\pi^2 T_\varepsilon T_\sigma V^2) \left(\cos\frac{2\pi x}{l} \right. \right. \\
& \left. - e^{\left(x-a_j\right)/T_\sigma V} \cos\frac{2\pi a}{l} \right) - 2\pi l V (T_\varepsilon - T_\sigma) \left(\sin\frac{2\pi x}{l} - e^{\left(x-a_j\right)/T_\sigma V} \sin\frac{2\pi a}{l} \right) \\
& \left. + 2D_j l^2 h \left(1 - e^{\left(x-a_j\right)/T_\sigma V}\right) \right] + \frac{E}{2H} \left(2D_j - h_j\right) \left(1 - e^{\left(x-a_j\right)/T_\sigma V}\right).
\end{aligned}
\tag{5.94}
$$

A similar expression is obtained for the unknown displacement function $u(x, y_j)$ in the adhesion interval $b_j < x < l - a_j$.

The condition of existence of the regime of discrete contact with saturated adhesion follows from the adopted model of adhesion given by Eqs. (5.85) and (5.86). The maximum value of the gap between the surfaces

$u(x, y_j) - f(x, y_j) - D_j$ must not exceed the radius of adhesion force action h_0:

$$\max\left(u(x, y_j) - f(x, y_j) - D_j\right) \leq w_a/p_0. \tag{5.95}$$

If the function $u(x, y_j)$ in the interval $b_j < x < l - a_j$ does not satisfy Eq. (5.95), then the regime of discrete contact with saturated adhesion cannot occur in the j-strip. In this case, the gap is too large, and discrete zones of adhesion can be formed around contact zones.

Discrete contact with discrete adhesion (Fig. 5.25c). In this case, contact occurs in the interval $\Omega_j^c = \left(-a_j, b_j\right)$, and the adhesion pressure $-p_0$ acts in the intervals $\Omega^a{}_j = \left(-a_{1j}, -a_j\right) \cup \left(b_j, b_{1j}\right)$. The intervals $\left(-l/2, -a_{1j}\right) \cup \left(b_{1j}, l/2\right)$ are free of loading. So, different boundary conditions take place for three different regions. The differential Eq. (5.84) is solved in the interval $-a_j < x < b_j$ for the contact pressure $p(x, y_j)$ and in the remaining intervals of the cell of periodicity for the displacement $u(x, y_j)$. For the determination of the unknown constants, the continuity conditions for the pressure $p(x, y_j)$ and displacement $u(x, y_j)$ are stated at the points $x = -a_{1j}$, $x = -a_j$ and $x = b_j$, $x = b_{1j}$, as well as the conditions of periodicity. To determine the end points of the adhesion regions a_{1j} and b_{1j}, we use the conditions following from the adopted model of adhesion given by Eqs. (5.85 and 5.86), according to which the gap between the surfaces at the points $x = -a_{1j}$ and $x = b_{1j}$ must equal h_0:

$$\begin{aligned} u\left(-a_{1j}, y_j\right) - f\left(-a_{1j}, y_j\right) - D_j &= w_a/p_0 \\ u\left(b_{1j}, y_j\right) - f\left(b_{1j}, y_j\right) - D_j &= w_a/p_0. \end{aligned} \tag{5.96}$$

As a result, we have four nonlinear algebraic equations to be numerically solved for a_j, b_j, a_{1j}, and b_{1j}, under the condition that the penetration D_j is given in the j-th strip. Then the pressure $p(x, y_j)$ in the contact region $-a_j < x < b_j$ is determined by Eq. (5.94).

If those four equations cannot be solved for a_j, b_j, a_{1j}, and b_{1j} so that the roots are positive and do not exceed $l/2$, then we can assume that the surfaces are not in contact with each other. In this case, one of the following contactless regimes occurs: adhesion acting over all surface, adhesion acting over discrete regions, or no interaction between the surfaces. Solutions for these regimes and conditions of their existence are obtained in a similar way to the solutions of the contact regimes.

After solving the contact problem for each strip, normal and tangential force, acting on each cell of periodicity, and the coefficient of friction μ are calculated from the relations

$$P = 2\Delta \sum_{j=1}^{N} \int_{-l/2}^{l/2} p_j(x)\, dx,\ T = 2\Delta \sum_{j=1}^{N} \int_{-l/2}^{l/2} \tau_j(x)\, dx,\ \mu = T/P. \quad (5.97)$$

Where the tangential stress $\tau_j(x)$ is defined by the geometrical relation

$$\tau(x, y_j) = p(x, y_j)\sin\left[\operatorname{arctg}\left(f'(x, y_j)\right)\right] \approx p(x, y_j) f'(x, y_j), \quad (5.98)$$

which is valid under the condition that $l << h$.

5.6.3 Effect of the parameters of adhesion and surface relief on the contact characteristics and friction force

For the calculation, we use the following dimensionless values: contact pressure pR/γ, coordinates x/l, y/l, normal penetration of an asperity D/h into the viscoelastic layer, normal load on an asperity $P/(l\gamma)$, friction force on an asperity $T/(l\gamma)$. The results obtained depend on the dimensionless viscosity parameter T_ε/T_σ, adhesion parameter $\lambda_w = p_0(k/w_a)^{1/2}$, where $k = H/E$, velocity parameter $l/(VT_\varepsilon)$, and the waviness parameter h/l. The parameter λ_w is the analog of the adhesion parameter λ for the case of the Winkler-type 1-D foundation (see Section 4.1.3).

The calculations were carried out to compare the results for sliding of a separate spherical asperity and a wavy surface, and also to analyze the influence of the surface geometry, adhesion, and sliding velocity on the contact pressure distribution, areas of contact and adhesive interaction, and deformation component of the friction force. For the calculations, the radius R of a separate asperity is taken to be equal to the radius of a tip of the wavy surface asperity $R = l^2/(2\pi^2 H)$.

In Fig. 5.27, the distributions of the dimensionless contact pressure pR/γ are presented for a separate spherical asperity (a) and for the wavy surface (b) as the same dimensionless normal load is applied to an asperity $P/(R\gamma) = 400$. The distribution graph for the wavy surface (b) is presented on the half of a periodicity cell $\frac{x}{l} \in [-0.5; 0.5]$, $\frac{y}{l} \in [-0.5; 0]$, and the graph for the separate asperity (a) is presented on the same domain for the

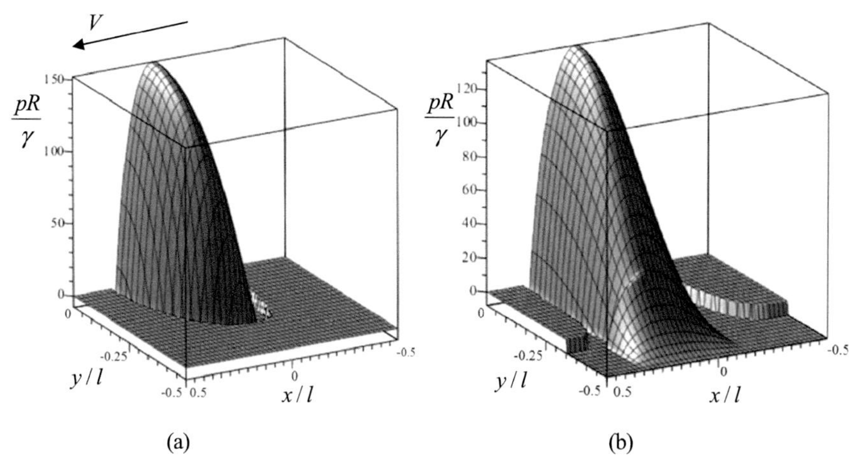

Figure 5.27 Contact pressure distributions for a single spherical asperity (a) and for an asperity of the wavy surface (b).

convenience of comparison. The values of the parameters used for calculation are $T_\varepsilon/T_\sigma = 10$, $\lambda_w = 1$, $l/(VT_\varepsilon) = 0.4$, and $l/h = 4$. The results indicate that viscoelastic properties of the foundation lead to a assymmetrical pressure distribution (shifted in the direction of sliding). Adhesion leads to the negative constant pressure $p(x, y) = -p_0$ acting in an area outside the contact region. For the case of a wavy surface, the maximum contact pressure is smaller, while areas of contact and adhesive interaction are wider than for the separate asperity.

For small loads, for which only tips of the wavy surface are in contact with the viscoelastic foundation, the contact pressure distributions are similar to those for the separate asperity. As the load per asperity increases and the interaction becomes closer to the saturated contact, the difference of the results for two models significantly increases. Fig. 5.28 shows the areas of contact (dark-grey regions) and adhesion (light-grey regions) for the separate asperity, (a) and (b), and for the wavy surface, (c) and (d). The graphs (a) and (c) correspond to the dimensionless load $P/(R\gamma) = 200$ for which we have separate contact spots for both cases, only slightly different from each other. The graphs (b) and (d) correspond to the load $P/(R\gamma) = 550$, at which contact spots begin to coalescence for the wavy surface (d) and the contact and adhesion areas are significantly different in size and shape from those for one asperity (b).

Fig. 5.29 illustrates the effect of the contact density (distance between asperities) on the contact characteristics. The dimensionless nominal

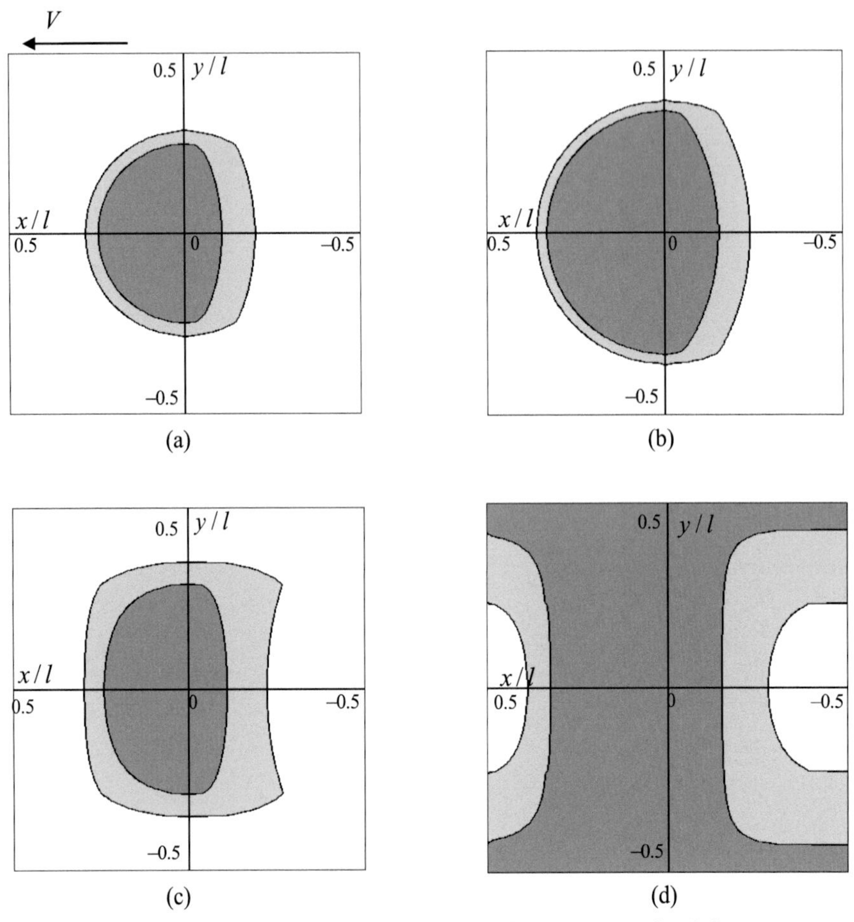

Figure 5.28 Contact and adhesion areas under the increasing load for a separate asperity (a), (b) and for a wavy surface (c), (d).

pressure $Ph/l^2\gamma$ and tangential stress $(Th/l^2\gamma)$ are calculated for two values of the dimensional distance between asperities of the wavy surface — $l/h = 3$ (curves 1 and 1′) and $l/h = 6$ (curves 2 and 2′). Results for solid lines 1 and 2 are calculated for the following parameters: $T_\varepsilon/T_\sigma = 10$, $l/(VT_\varepsilon) = 0.4$, $\lambda_w = 15$. Results for dashed lines 1′ and 2′ are calculated for the same parameters but with no adhesion taken into account: $\lambda_w = 0$. Note that the value of the adhesion parameter $\lambda = 15$, for which curves 1 and 2 are obtained, is much higher than that for the results presented in Figs. 5.27 and 5.28, so the effects of adhesion for the results presented in Fig. 5.58 are more significant. Higher values of the adhesion parameter λ_w

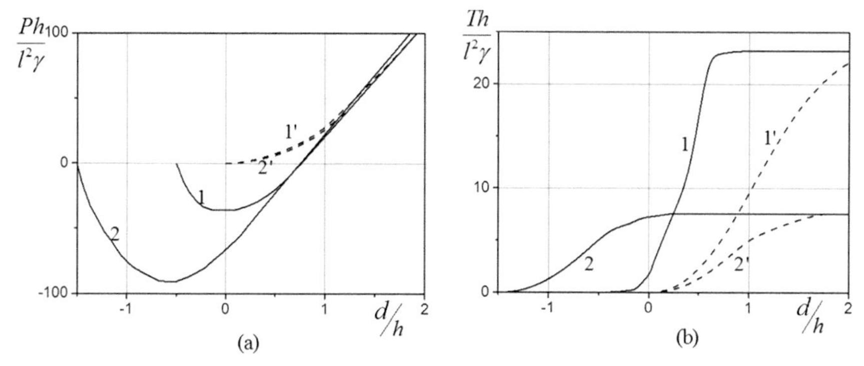

Figure 5.29 Nominal normal pressure (a) and tangential stress (b) versus normal penetration for two different dimensionless distances between asperities of the wavy surface with adhesion (solid lines) and no adhesion (dashed lines).

correspond to softer material with higher surface energy. In this case, the contact exists also for considerably negative loads and penetrations. The larger the distance is between asperities, the higher is the negative load at which contact between surfaces exists. The dependence of the load on the penetration is nonmonotone, as is the case for the elastic adhesive contact (Chapter 4). At high penetrations, the contact becomes saturated and the friction force attains a constant value. This value is higher for smaller distance between asperities. Therefore, decreasing the distance between asperities (increasing contact density) leads to smaller effect of adhesion but higher effect of hysteretic losses. Due to this, at negative and small values of the penetration, the friction force is higher for surfaces with larger distance between asperities, whereas for deep penetrations, the friction force for larger distance between asperities is lower.

In Fig. 5.30, the coefficient of friction $\mu = T/P$ is presented as a function of the external normal load applied to an asperity of the wavy surface. Curve 1 corresponds to the adhesion parameter $\lambda_w = 0$, curve 2 to $\lambda_w = 5$, curve 3 to $\lambda_w = 10$, and curve 4 is constructed for the limit case $\lambda_w \to \infty$. In the case of no adhesion (curve 1), the coefficient of friction is zero for zero load. As the load increases, the coefficient of friction increases, attains its maximum, and then decreases, which is accounted for by the viscous properties of material and effect of saturation of the contact region at high loads. The combined action of viscosity and adhesion leads to

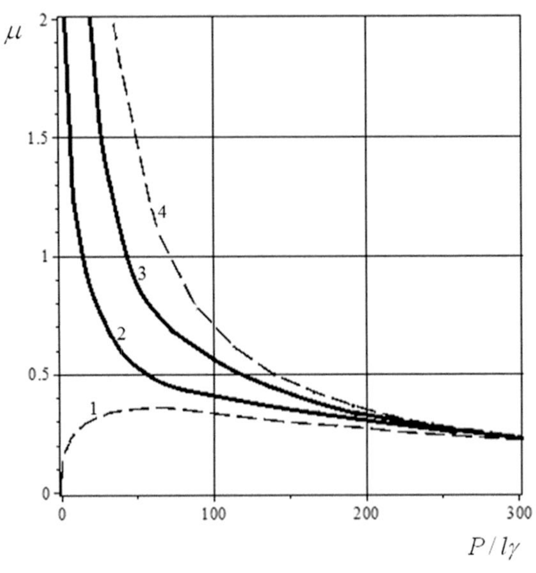

Figure 5.30 Coefficient of friction as a function of normal load in the case of no adhesion (curve 1) and for various values of the adhesion parameter (curves 2–4).

nonzero coefficient of friction at any positive load. The specific effect of adhesion is a sharp growth of the coefficient of friction at small loads.

Note that in the case where the adhesion attraction between the surfaces is strong enough so that the contact is saturated over the entire surface of the wavy indenter, from Eq. (5.97) taking into account Eqs. (5.92), (5.89), and (5.98), we obtain a simple analytic expression for the coefficient of friction:

$$\mu = \frac{\pi^2 l^2 h^2 V(T_\varepsilon - T_\sigma)E}{H(l^2 + 4\pi^2 T_\sigma{}^2 V^2)P}. \tag{5.99}$$

From Eq. (5.99) it follows that in the case of saturated contact, the coefficient of friction nonmonotonically depends on the velocity V and distance between asperities l, inversely proportional to the normal load P, and proportional to the squared height of asperities h. Note that in this case, the coefficient of friction does not depend on the adhesion parameter λ_w.

5.7 Friction of a multiscale wavy surface taking into account adhesion

Since the main feature of a rough surface is its multiscale geometry, in this section the model is presented to study this effect for a punch with multiscale wavy surface on the friction force in sliding over the viscoelastic base.

5.7.1 Model description

Consider sliding with the constant velocity V of a rigid indenter over the surface of the viscoelastic foundation. The surface of the rigid indenter is characterized by waviness at several scale levels described by a set of heights h_i of asperities and distances l_i between them, where $i = 1..M$, and$M$ is the number of the scale levels taken into account. Asperities of the i-th level are imposed on the surface of asperities of the $(i$-1)-th level (Fig. 5. 31).

Mechanical properties of the viscoelastic foundation are described by the Kelvin model (Eq. 5.49). The adhesion attraction of surfaces is described by the Maugis model given by Eqs. (5.85) and (5.86).

At each scale level, the friction force is calculated as a sum of two terms. The first term is due to hysteretic losses occurring when asperities of this

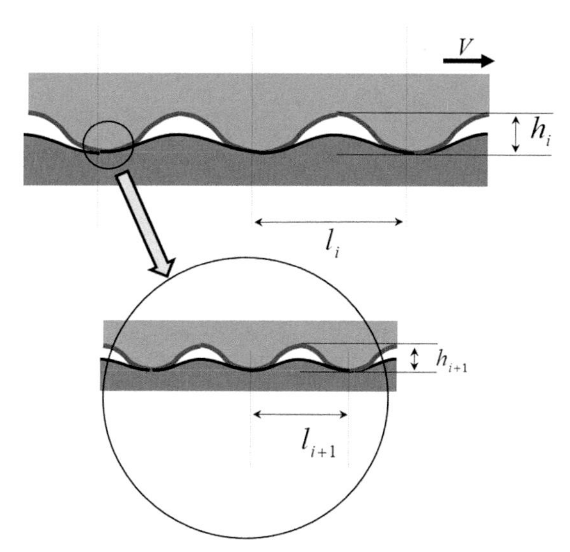

Figure 5.31 Scheme of contact between the multilevel wavy surface and a viscoelastic body at two levels.

scale level cyclically deform the viscoelastic foundation during sliding. The second term is the law of friction determined from the solution of the contact problem at the inferior scale level. This friction law accounts for the contribution of hysteresis and adhesion at all smaller scale levels (Makhovskaya, 2019).

5.7.2 Problem formulation and solution at *i*-th scale level

At an i-th scale level, the waviness of the surface is described by the function periodic in two directions (Fig. 5.16):

$$f_i(x,y)=h_i-\frac{h_i}{4}\left(\cos\left(\frac{2\pi x}{l_i}\right)+1\right)\left(\cos\left(\frac{2\pi y}{l_i}\right)+1\right). \tag{5.100}$$

We pass from the system of coordinates (x', y', z') attached to the viscoelastic base to the system of coordinates (x, y, z) moving with the wavy surface with the constant velocity V according to Eq. (5.71). In the moving system, the stress and displacement do not depend explicitly on time, and Eq. (5.49) for the i-th scale level takes the form:

$$u_i - VT_\varepsilon\frac{\partial u_i}{\partial x}=\frac{H}{E}\left(p_i - T_\sigma V\frac{\partial p_i}{\partial x}\right). \tag{5.101}$$

The boundary conditions for the stress and displacement of the viscoelastic foundation are the following:

$$\begin{aligned} u_i(x,y) &= f_i(x,y)+D_i, && (x,y)\in\Omega_i^c \\ p_i(x,y) &= -p_0, && (x,y)\in\Omega_i^a \\ p_i(x,y) &= 0, && (x,y)\notin\Omega_i^c\cup\Omega_i^a \end{aligned}, \tag{5.102}$$

where Ω_i^c and Ω_i^a are the contact and adhesion regions, respectively, and D_i is the maximum penetration of the wavy surface into the foundation at the i-th level. The equilibrium condition must be satisfied for the nominal pressure $\overline{p}_i$:

$$\overline{p}_i=\frac{1}{l_i^2}\iint\limits_{\Omega_i} p_i(x,y)\,dxdy, \tag{5.103}$$

where $\Omega_i=\Omega_i^c\cup\Omega_i^a$ is the region of interaction at the i-th level.

The problem at the i-th scale level is solved by using the strip method. A cell of periodicity $x\in(-l_i/2, l_i/2)$, $y\in(-l_i/2, l_i/2)$ is divided into $2N$

strips of a thickness Δ parallel to the x-axis (Fig. 5.25). Penetration of the indenter into the j-th strip at the i-th scale level D_i^j is given by the relation:

$$D_i^j = \frac{h_i}{2}\left(\cos\left(\frac{2\pi y_j}{l_i}\right) - 1\right) + D_i. \tag{5.104}$$

So, the 3-D contact problem is reduced to a 2-D problem for each strip. The differential Eq. (5.101) with conditions Eq. (5.102), where D_i^j given by (5.104) is substituted instead of D_i, is solved analytically to obtain the closed form relation for the normal stress distribution. The regime of interaction in the gap should be appropriately chosen for each strip depending on the value of D_i^j: saturated contact, partial contact with saturated adhesion, partial contact with discrete adhesion, or contactless regimes. Description of the method of solution of the 2-D contact problem and of the strip method is given in Section 5.6.2.

As a result, the function of contact stress $p_i(x, y_j)$ is constructed analytically for each strip. The coordinates of the boundaries $-a_i^j$, b_i^j of the contact region and $-a_{1i}^j$, b_{1i}^j of the adhesion region are calculated numerically in each strip. After this, the contact area can be calculated:

$$A_i = 2\Delta \sum_{j=1}^{N} \left(a_i^j + b_i^j\right). \tag{5.105}$$

When the distribution of the normal stress $p_i(x, y_j)$ is known for a j-th strip, the tangential stress can be calculated in accordance with the relation:

$$\tau_i\left(x, y_j\right) = p_i\left(x, y_j\right) \frac{\partial f_i\left(x, y_j\right)}{\partial x}. \tag{5.106}$$

By averaging the obtained normal and tangential stresses over the cell of periodicity, we can calculate the mean (nominal) normal and tangential stress at the i-th scale level:

$$\overline{p}_i = \frac{2\Delta}{l_i^2} \sum_{j=1}^{N} \int_{-l_i/2}^{l_i/2} p_i\left(x, y_j\right) dx, \overline{\tau}_i = \frac{2\Delta}{l_i^2} \sum_{j=1}^{N} \int_{-l_i/2}^{l_i/2} \tau_i\left(x, y_j\right) dx. \tag{5.107}$$

In Fig. 5.32, calculated results for an isolated scale level are presented. The ratio of the contact area A_i to the full contact area $A_{\text{full}} = l_i^2$ as a function of nominal stress is shown in Fig. 5.32a. The mean frictional stress

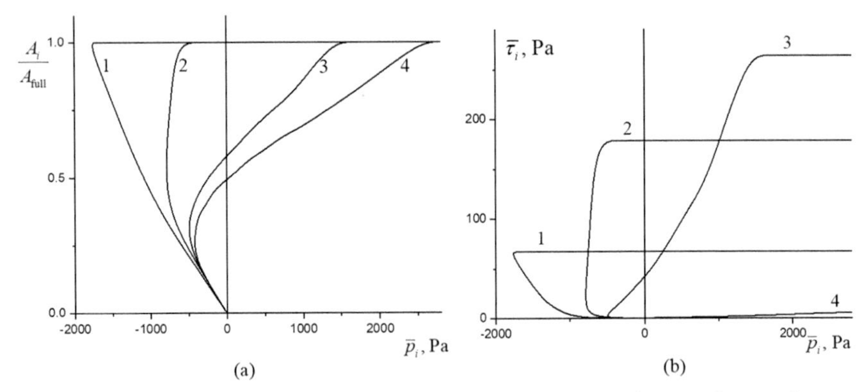

Figure 5.32 Contact area ratio versus nominal pressure (a) and mean frictional stress versus nominal pressure (b) at a separate scale level for various sliding velocities: $V = 0.0001$ m/s (curves 1), $V = 0.0003$ m/s (curves 2), $V = 0.001$ m/s (curves 3), $V = 0.08$ m/s (curves 4).

as a function of nominal stress is presented in Fig. 5.32b. Calculations were carried out in the dimensional form for the geometric characteristics of the waviness $l_i = 500$ nm and $h_i = 100$ nm, mechanical properties of the viscoelastic foundation $E/H = 1$ GPa/m, $T_\varepsilon = 0.01$ s, $T_\varepsilon/T_\sigma = 100$, and the adhesion parameters $w_a = 2.5 \times 10^{-5}$ N/m, $h_0 = 10$ nm. The results are presented for various values of the sliding velocity V.

From the results presented in Fig. 5.32, it is seen that both the contact area and frictional stress attain a constant value at some nominal pressure and remain constant when the pressure further increases. It is accounted for by the saturation of contact when the gap is zero over all the contact surface. Taking into account adhesion attraction leads to the existence of contact at negative nominal pressures, which is more pronounced at smaller velocities. Note that the hysteresis friction force is positive (directed against the sliding direction) both for positive and for negative nominal stress.

The graphs presented in Fig. 5.32b are ambiguous for some values of the velocity, e.g., for $V = 0.1$ mm/s. This ambiguity is caused by the combined effect of compliance and adhesion, and it leads to a hysteresis in cyclic normal approach and separation of the surfaces, which is also the case for purely elastic bodies (see Chapter 3).

The function $\overline{\tau}_i = \overline{\tau}_i[\overline{p}_i]$ constructed for the i-th level (examples of which are presented in Fig. 5.32b) is used as a law of friction at the superior $(i-1)$-th scale level (see Fig. 5.1). In the domain of ambiguity of the function $\overline{\tau}_i = \overline{\tau}_i[\overline{p}_i]$, the upper part of the curve is used for calculation, which corresponds to the stable solution at unloading.

5.7.3 Constructing the solution for a multilevel wavy surface

To construct the solution for a M-scaled wavy indenter, the problem is first solved at the smallest M-th scale. From this solution, the law of friction $\overline{\tau}_M = \overline{\tau}_M[\overline{p}_M]$ is constructed numerically by using relations Eqs. (5.106) and (5.107) for $i = M$. For the $(M-1)$-th level, the tangential stress is a sum of two terms. The first term is due to hysteresis occurring when asperities of this scale level cyclically deform the viscoelastic foundation during sliding. The second term follows from the friction law determined at the M-th level, which is defined at each point by the normal stress $p_{M-1}(x, y_j)$:

$$\tau_{M-1} = p_{M-1}(x, y_j)\frac{\partial}{\partial x}f_{M-1}(x, y_j) + \overline{\tau}_M\left[p_{M-1}(x, y_j)\right]. \tag{5.108}$$

By averaging Eq. (5.108) over the cell of periodicity $x \in (-l_i/2, l_i/2)$, $y \in (-l_i/2, l_i/2)$, we calculate the mean frictional stress of the $(M-1)$-th level:

$$\begin{aligned}\overline{\tau}_{M-1} = &\frac{2\Delta}{l_{M-1}^2}\sum_{j=1}^{N}\int_{-l_{M-1}/2}^{l_{M-1}/2} p_{M-1}(x, y_j)\frac{\partial}{\partial x}f_{M-1}(x, y_j)dx \\ &+\frac{2\Delta}{l_{M-1}^2}\sum_{j=1}^{N}\int_{-l_{M-1}/2}^{l_{M-1}/2} \overline{\tau}_M\left[p_{M-1}(x, y_j)\right]dx,\end{aligned} \tag{5.109}$$

whose dependence on the nominal pressure $\overline{p}_{M-1}$ will in turn serve as a law of friction for the $(M-2)$-th level: $\overline{\tau}_{M-1} = \overline{\tau}_{M-1}[\overline{p}_{M-1}]$.

By subsequently repeating this procedure, one can calculate the total coefficient of friction for the multilevel wavy surface:

$$\begin{aligned}\mu = \frac{\overline{\tau}_1}{\overline{p}_1} = &\frac{2\Delta}{\overline{p}_1 l_1^2}\sum_{j=1}^{N}\int_{-l_1/2}^{l_1/2} p_1(x, y_j)\frac{\partial}{\partial x}f_1(x, y_j)dx + \frac{2\Delta}{\overline{p}_1 l_1^2}\sum_{j=1}^{N} \\ &\int_{-l_1/2}^{l_1/2} \overline{\tau}_2\left[p_1(x, y_j)\right]dx.\end{aligned} \tag{5.110}$$

The first term in the right-hand side of Eq. (5.110) is the contribution of asperities of the first (largest) scale level into the hysteresis friction force. The second term is the contribution of the second and all the subsequent scale levels up to the M-th one.

The solution of the multilevel problem requires the contact problem to be subsequently solved at each scale level. The most difficult part of the solution is the determination of the boundaries $-a_i^j$, b_i^j of the contact region and the boundaries $-c_i^j$, d_i^j of the adhesion region in a cell of periodicity for each scale level i and each number j of the strip, which includes the determination of the regime of filling of the gap in each strip. These values are calculated by numerically solving a system of two to four algebraic equations.

In the particular case, where the external load and adhesion are high enough so that the saturated contact occurs at all scale levels, the solution is significantly simplified. The pressure distribution at each scale level is then specified by a simple analytic relation in Eq. (5.92) for each strip. By substituting Eq. (5.92) into Eqs. (5.106) and (5.107), the relation for the frictional stress $\overline{\tau}_i$ independent of the nominal pressure $\overline{p}_i$ is obtained. This allows one to calculate the total friction force by direct summation of contributions of all scale levels. As a result, the total coefficient of friction has the form:

$$\mu = \frac{\pi^2 V(T_\varepsilon - T_\sigma)E}{\overline{p}_1 H} \sum_{i=1}^{N} \frac{h_i}{4\pi^2 V^2 T_\sigma^2 + l_i^2}, \tag{5.111}$$

where $\overline{p}_1$ is the external nominal pressure, which coincides with the nominal pressure at the first scale level. The obtained expression Eq. (5.111) is the generalization of expression Eq. (5.99) to the case of a multiscale indenter.

5.7.4 Results of calculation

Calculation is carried out for a two-level surface with the waviness parameters $l_1 = 10$ μm, $h_1 = 1$ μm and $l_2 = 500$ nm, $h_2 = 100$ nm. The properties of the viscoelastic foundation and adhesion parameters are the same as for the results presented in Fig. 5.32. The external nominal pressure is constant and equal to 0.1 kPa, and the sliding velocity ranges from 0 to 0.1 m/s.

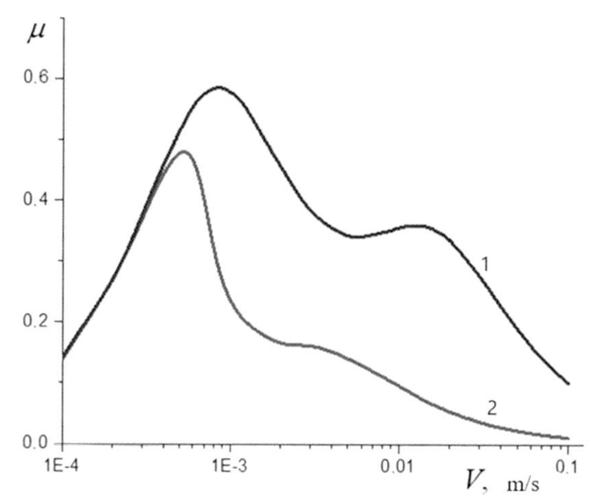

Figure 5.33 Coefficient of friction as a function of sliding velocity for two different works of adhesion in the case of a two-level wavy indenter: $\gamma = 10^{-4}$ N/m (curve 1) and $\gamma = 2.5{*}10^{-3}$ N/m (curve 2)

In Fig. 5.33, the total friction coefficient for the two-level surface as a function of the sliding velocity is presented for two values of the specific energy of adhesion. The right peak corresponds to the contribution of the first scale level, and the left peak to the second level.

The results indicate that depending on the values of waviness parameters, the coefficient of friction as a function of sliding velocity can have several peaks, each being associated with hysteretic losses at some scale level. It was first observed by Grosch that the master curve of rubber friction can have more than one peak (Grosch, 1963). He concluded that in the general case, the master curve has two peaks due to two mechanisms of rubber friction: adhesion and hysteretic losses. Master curves with two peaks were also obtained experimentally (Moore, 1980) and by numerical modeling (Nettingsmeier and Wriggers, 2004). In the present study, a phenomenological law of friction is obtained from considering the contribution of lower scale levels including both hysteretic losses and adhesion forces acting in the direction normal to the surface. In this case it is impossible to say that one peak is due to adhesion and the other is due to hysteresis, for both peaks in Fig. 5.33 are due to hysteretic resonance and both are strongly influenced by adhesion.

The graphs in Fig. 5.33 show also that increasing the specific energy of adhesion not only increases the coefficient of friction, but it also shifts the peaks in the direction of higher velocities.

The model presented takes into account not only partial contact but also partial regions of adhesion. When the contact is assumed complete at all scale levels, the calculation of the friction force is significantly simplified (see Eq. 5.111). In Fig. 5.34, the graph of the coefficient of friction as a function of velocity is presented for the case where the external nominal pressure is high enough to ensure the saturated contact at all three scale levels that are taken into account. These scale levels are $l_1 = 10$ μm, $h_1 = 1$ μm, $l_2 = 500$ nm, $h_2 = 100$ nm, and $l_3 = 5$ nm, $h_3 = 5$ nm. The nominal pressure is held constant, $\bar{p}_1 = 5$ MPa. All the remaining parameters coincide with those used for calculating the results presented in Fig. 5.33. Curves 1, 2, and 3 correspond to contributions of the first, second, and third scale levels, respectively. Curve 4 correspond to the total coefficient of friction, which is, in this case, a direct sum of the three components.

The results show that the peak associated with the largest (1st) scale level is the highest, whereas the 2nd and 3rd levels give significantly smaller peaks. By comparing the results presented in Fig. 5.33 with those of

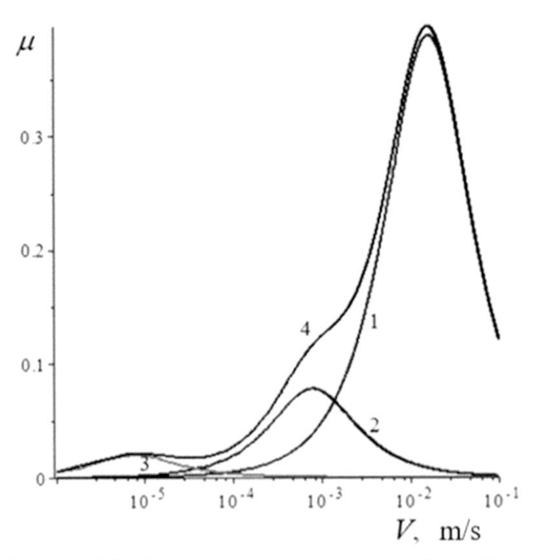

Figure 5.34 Coefficient of friction as a function of velocity (contribution of three scale levels) in the case of saturated contact.

Fig. 5. 34, one can conclude that neglecting partial character of contact and adhesion interaction can lead to substantial underestimation of the contribution of smaller scale levels into the total friction force.

5.8 Effect of fluid in the gap in sliding contact of a punch with periodic microgeometry over the viscoelastic half-space

In this section, the sliding contact of a punch with 2-D periodic surface microgeometry and the viscoelastic foundation is analyzed taking into account the existence of incompressible fluid in the gap. It allows us to analyze the combined effect of surface microgeometry, relaxation characteristics of the base, as well as volume of the fluid on the friction force and contact pressure distribution.

5.8.1 Problem formulation

Consider a rigid punch, whose shape of the surface is described by the doubly periodic function (Fig. 5.35):

$$f(x,y) = h\left(1 - \frac{1}{4}\left(1+\cos\frac{2\pi x}{l_1}\right)\left(1+\cos\frac{2\pi y}{l_2}\right)\right), \tag{5.112}$$

where l_1 is the period in the x-direction, l_2 is the period in the y-direction, and h is the height of asperities. The punch is acted on by the normal

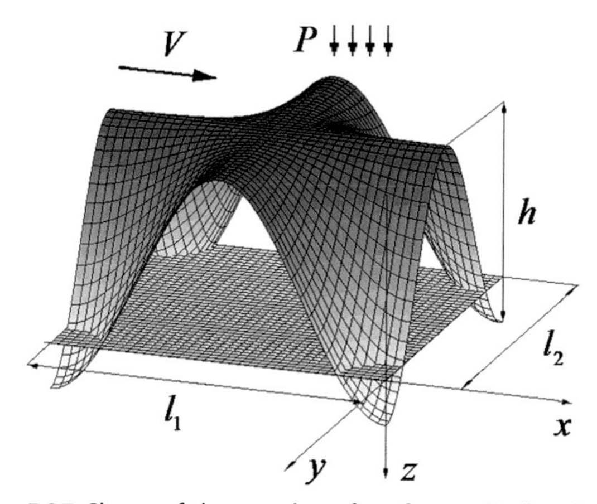

Figure 5.35 Shape of the punch surface for a cell of periodicity.

external force P on each asperity and slides with the velocity V in the x-direction over the boundary of a viscoelastic layer of thickness H bonded to the rigid base.

There is incompressible fluid in the gap between the surfaces, and its volume v is prescribed for a cell of periodicity so that

$$0 \le v \le v_{\max}; \quad v_{\max} = \int_0^{l_2}\int_0^{l_1} f(x,y)\,dx\,dy. \tag{5.113}$$

The relation between the normal displacement $u = u(x,y)$ of the upper boundary of the viscoelastic layer and the normal pressure $p = p(x,y)$ is described by the Kelvin model given by Eq. (5.49). In the moving system of coordinates attached to the punch, Eq. (5.49) has the form (5.84).

The boundary conditions on the upper surface of the viscoelastic layer have the following form. In the contact region Ω between the punch and the layer, the tangential stresses are zero, while for the normal displacement the contact condition is satisfied:

$$u = f(x,y) + D. \tag{5.114}$$

In the regions occupied by fluid, we have

$$p(x,y) = p_f, \quad (x,y) \notin \Omega, \tag{5.115}$$

where p_f is the hydrostatic pressure in fluid that depends on the external load applied to the punch. The equilibrium condition is satisfied:

$$P = \int_{-y_c}^{y_c}\int_{-a(y)}^{b(y)} p(x,y)\,dx\,dy + p_f\left(l_1 l_2 - \int_{-y_c}^{y_c}\int_{-a(y)}^{b(y)} dx\,dy\right), \tag{5.116}$$

where $-a(y)$ and $b(y)$ are the unknown functions defining the end points of the contact region Ω in the (x,y) plane, which are unknown, $2y_c$ is the maximum width of the contact area in the y-direction.

The tangential force arising due to asymmetric distribution of the contact pressure (hysteretic friction force) is given by the relation:

$$T = \int_{-y_c}^{y_c}\left(\int_{-a(y)}^{b(y)} f_x(x,y)p(x,y)\,dx + \int_{-b(y)}^{-a(y)} f_x(x,y)p_f\,dx\right)dy, f_x = \partial f(x,y)/\partial x. \tag{5.117}$$

Unknown constants—the punch indentation D and fluid pressure p_f—are calculated from the condition of equilibrium Eq. (5.115) and the condition of constant volume v of fluid in each periodicity cell:

$$v = \frac{3h}{4} - 2\int_0^{1/2}\int_0^1 (d - u_z(x, y))dxdy. \tag{5.118}$$

To solve the problem stated above, the strip method was used (Goryacheva and Shpenev, 2012) and the algorithm of calculation was similar to that presented in Section 5.6.2.

5.8.2 Analysis of the computational results

The problem solution was analyzed depending on the following dimensionless parameters: parameter $\overline{v} = \frac{v}{L_1 L_2 H}$ defined by the fluid volume per one period, dimensionless height of asperities $\overline{h} = \frac{h}{H}$; dimensionless load acting on each asperity $\overline{P} = \frac{P}{El_1 l_2}$; and the parameters $\overline{T_\varepsilon} = \frac{T_\varepsilon V}{l_1}$, $\overline{T_\sigma} = \frac{T_\sigma V}{l_1}$ relating the relaxation and retardation times of the viscoelastic material to the sliding velocity.

In Fig. 5.36, a dashed line represents the dependence of gap volume in a cell of periodicity in the absence of fluid ($p_f = 0$, $h = 2$) on the parameter $\overline{T_\varepsilon}$, which is proportional to the punch velocity. As the velocity increases, the volume of the gap increases monotonically, and it has a horizontal asymptote as the velocity tends to infinity. From this it follows that if the

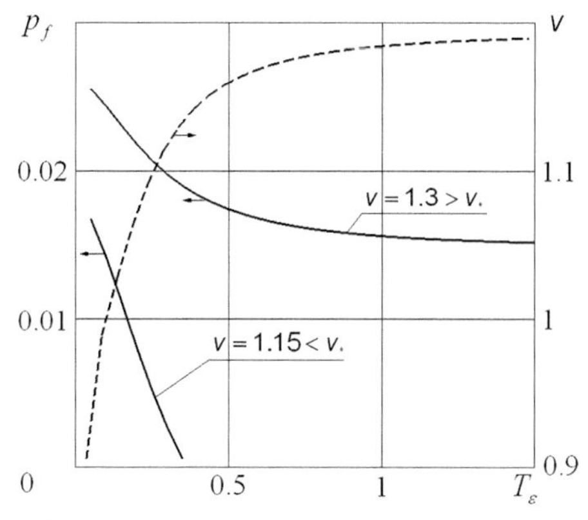

Figure 5.36 Hydrostatic pressure in fluid (solid lines) and fluid volume for a cell of periodicity (dashed line) as a function of the parameter $\overline{T_\varepsilon}$.

fluid volume is less than a critical value $v* = \lim_{\overline{T_\varepsilon} \to \infty} v\big|_{p_f=0}$, then as the velocity increases, at some instant the fluid pressure under the punch falls to zero, and fluid has no effect on the process of sliding any more. If the fluid volume under the punch exceeds the critical value, then fluid influences the process of sliding at any velocities.

This is illustrated by solid lines in Fig. 5.36 that show the fluid pressure as a function of the velocity parameter. For the fluid volume smaller than the critical value, the fluid pressure drops to zero. For higher fluid volumes, the fluid pressure tends to a constant value as the velocity tends to infinity. In the calculations, the following values of the parameters were used: $\overline{P} = 0.03$, $\overline{T_\sigma} = 0.13\overline{T_\varepsilon}$.

The shape of contact area is shown in Fig. 5.37 in the presence of fluid (solid lines) and in the absence of fluid (dashed lines) for various values of the parameter $\overline{T_\varepsilon}$. Presence of fluid under the punch decreases the size of the contact area. If the fluid volume exceeds the critical value, presence of fluid makes the contact area more symmetric.

The hysteretic component of the coefficient of friction $\mu = T/P$, calculated by Eqs. (5.116) and (5.117), as a function of the parameter $\overline{T_\varepsilon}$ is presented in Fig. 5.38 for various values of fluid volume in the gap. The dashed line corresponds to no fluid in the gap. These functions are nonmonotonic with a maximum value. Presence of fluid in the gap decreases the coefficient of friction and its maximum value.

The model presented can be used for various application, e.g., for prediction of the aquaplaning effect in contact between a tire thread and wet asphalt. From the results obtained it follows that presence of fluid in the gap leads to a decrease in the contact area and hysteretic friction force. If the

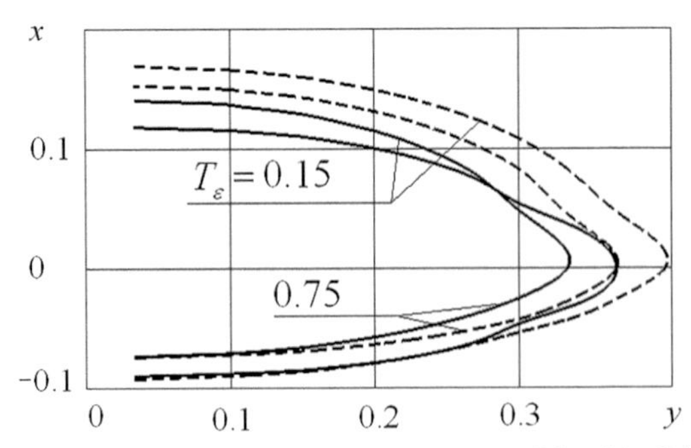

Figure 5.37 Shape of the contact area in the presence of fluid (solid lines) and in the absence of fluid (dashed lines) for various values of the parameter $\overline{T_\varepsilon}$.

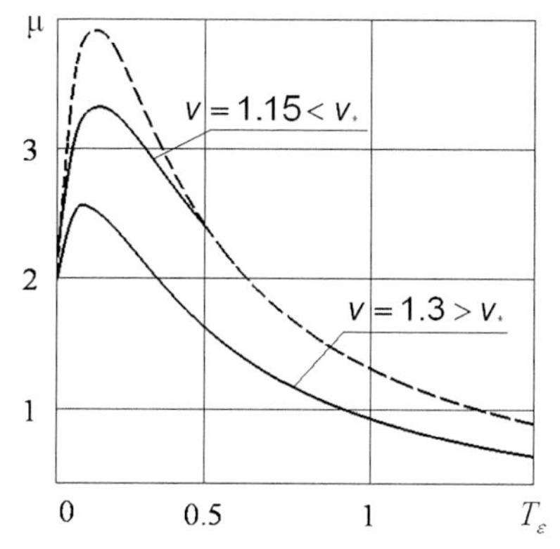

Figure 5.38 Hysteretic component of the coefficient of friction as a function of the parameter $\overline{T_\varepsilon}$ for various volumes of fluid in the gap.

fluid volume does not exceed the critical value, this effect vanishes as the velocity increases. If the fluid volume exceeds the critical value, the effect occurs at any sliding velocities.

5.9 Conclusions

The studies carried out in this chapter make it possible to evaluate the influence of surface microgeometry parameters on the contact characteristics (distribution of contact pressures, real contact area, and hysteretic component of the friction force) in friction of lubricated surfaces when tangential stresses in the contact interaction region can be neglected. As a result of modeling, new mechanical effects associated with the arrangement of asperities were described. In particular, for a doubly periodic shape of the surface relief, the possibility of merging of contact regions in the direction perpendicular to the direction of sliding was established for some sliding velocities.

The models developed allow one also to evaluate the influence of the adhesion forces on the contact characteristics in friction of viscoelastic materials against counterbodies that have a regular surface relief at one or several scale levels. It is shown, in particular, that taking into account adhesion can lead to a significant increase in the real contact area and hysteretic friction force. The transition from discrete to saturated contact occurs at lower loads than in the absence of adhesion. The contribution of different scale levels to the total friction force is analyzed depending on the sliding velocity and surface energy of interacting bodies.

It is shown that the presence of fluid in the gap between rough surfaces can lead to a decrease in the real contact area and in the coefficient of friction in a certain range of the sliding velocities.

The results obtained can be used to analyze the contact characteristics of elastomers in frictional interaction. Determination of the range of parameters, in which the hysteretic component of the friction coefficient is maximum, is of great importance for practical applications, for example, when calculating the friction of tires on the road surface.

References

Goryacheva, I.G., 1998. Contact Mechanics in Tribology. Kluwer Acad. Publ., p. 344

Goryacheva, I.G., Goryachev, A.P., 2016. Contact problems of the sliding of a punch with a periodic relief on a viscoelastic half-plane. J. Appl. Math. Mech. 80 (1), 73–83.

Goryacheva, I.G., Goryachev, A.P., 2020. Friction characteristic calculations during cylinder's sliding contact over the wavy viscoelastic base. J. Frict. Wear 41 (6), 502–508.

Goryacheva, I.G., Makhovskaya, Y.Y., 1997. Influence of surface layer imperfect elasticity on contact characteristics in sliding of rough elastic bodies. J. Frict. Wear 18 (1), 5–12.

Goryacheva, I.G., Makhovskaya, Y.Y., 2010. Modeling of friction at different scale levels. Mech. Solid. 45 (3), 390–398.

Goryacheva, I.G., Makhovskaya, Y.Y., 2015. Sliding of a wavy indenter on a viscoelastic layer surface in the case of adhesion. Mech. Solid. 50 (4), 439–450.

Goryacheva, I.G., Makhovskaya, Y.Y., 2016. Adhesion effect in sliding of a periodic surface and an individual indenter upon a viscoelastic base. J. Strain Anal. Eng. Des. 51 (4), 286–293.

Goryacheva, I.G., Sadeghi, F., 1995. Contact characteristics of rolling/sliding cylinder and a viscoelastic layer bonded to an elastic substrate. Wear 184, 125–132.

Goryacheva, I.G., Sadeghi, F., Nickel, D., 1996. Internal stresses in contact of rough body and a viscoelastic layered semi-infinite plane. ASME J. Tribol. 118 (1), 131–136.

Goryacheva, I.G., Shpenev, A.G., 2012. Modelling of a punch with a regular base relief sliding along a viscoelastic foundation with a liquid lubricant. J. Appl. Math. Mech. 76 (5), 582–589.

Greenwood, J.A., Tabor, D., 1958. The friction of hard sliders on lubricated rubber: the importance of deformation losses. Proc. Phys. Soc. 71 (6), 989.

Grosch, K.A., 1963. The relation between the friction and visco-elastic properties of rubber. Proc. R. Soc. Lond. Ser. A Math. Phys. 274 (1356), 21–39.

Hunter, S.C., 1961. The rolling contact of a rigid cylinder with a viscoelastic half space. Trans. ASME. Ser. E. J. Appl. Mech. 28, 611–617.

Kuznetsov, Y.A., 1985. Effect of fluid lubricant on the contact characteristics of rough elastic bodies in compression. Wear 102 (3), 177–194.

Makhovskaya, Y., 2019. Modeling sliding friction of a multiscale wavy surface over a viscoelastic foundation taking into account adhesion. Lubricants 7 (13).

Menga, N., Putignano, C., Carbone, G., Demelio, G.P., 2014. The sliding contact of a rigid wavy surface with a viscoelastic half-space. Proc. R. Soc. A. 470 (2169), 20140392.

Moor, D.F., 1980. Friction and wear in rubbers and tyres. Wear 61, 273–282.

Morozov, A.V., Makhovskaya, Y.Y., 2019. Effect of adhesion properties of frost-resistant rubbers on sliding friction. In: Radionov, A.A., Kravchenko, O.A., Guzeev, V.I., Rozhdestvenskiy, Y.V. (Eds.), Proceedings of the 4th International Conference on Industrial Engineering. Springer, pp. 1029–1037.
Nettingsmeier, J., Wriggers, P., 2004. Frictional contact of elastomer materials on rough rigid surfaces. PAMM Proc. Appl. Math. Mech. 4, 360–361.
Sheptunov, B.V., Goryacheva, I.G., Nozdrin, M.A., 2013. Contact problem of die regular relief motion over viscoelastic base. J. Frict. Wear 34, 83–91.
Staierman, I.Y., 1949. Contact Problem of the Elasticity Theory. (I.R.) Gostekhizdat, Moscow.

Index

Note: 'Page numbers followed by 'f' indicate figures those followed by 't' indicate tables and 'b' indicate boxes.'

W

Printed in the United States
by Baker & Taylor Publisher Services